48

99BB

15-

A CRITICAL PSYCHOLOGY
Interpretation of the Personal World

PATH IN PSYCHOLOGY
Published in Cooperation with Publications for the
Advancement of Theory and History in Psychology (PATH)

Series Editors:
David Bakan, *York University*
John Broughton, *Teachers College, Columbia University*
Miriam Lewin, *Manhattanville College*
Robert Rieber, *John Jay College, CUNY, and Columbia University*
Howard Gruber, *Rutgers University*

WILHELM WUNDT AND THE MAKING OF A SCIENTIFIC PSYCHOLOGY
Edited by R. W. Rieber

HUMANISTIC PSYCHOLOGY: Concepts and Criticisms
Edited by Joseph R. Royce and Leendert P. Mos

PSYCHOSOCIAL THEORIES OF THE SELF
Edited by Benjamin Lee

DEVELOPMENTAL APPROACHES TO THE SELF
Edited by Benjamin Lee and Gil G. Noam

THEORETICAL PSYCHOLOGY: The Meeting of East and West
A. C. Paranjpe

A CRITICAL PSYCHOLOGY: Interpretation of the Personal World
Edmund V. Sullivan

A CRITICAL PSYCHOLOGY
Interpretation of the Personal World

EDMUND V. SULLIVAN

Ontario Institute for Studies in Education
University of Toronto
Toronto, Ontario, Canada

PLENUM PRESS • NEW YORK AND LONDON

Library of Congress Cataloging in Publication Data

Sullivan, Edmund V.
   A critical psychology.

   (PATH in psychology)
   Bibliography: p.
   Includes index.
   1. Psychology—Methodology. 2. Social psychology. 3. Dialectic. I. Title. II. Series.
BF38.5.S83   1984   150'.72   83-20184
ISBN 0-306-41434-1

©1984 Plenum Press, New York
A Division of Plenum Publishing Corporation
233 Spring Street, New York, N.Y. 10013

Printed in the United States of America

To my children:

JEREMY, MEARA BRIGHID, AND DAMIAN

PREFACE

If the reader will excuse a brief anecdote from my own intellectual history, I would like to use it as an introduction to this book.

In 1957, I was a sophomore at an undergraduate liberal arts college majoring in medieval history. This was the year that we were receiving our first introduction to courses in philosophy, and I took to this study with a passion. In pursuing philosophy, I discovered the area called "philosophical psychology," which was a Thomistic category of inquiry. For me, "philosophical psychology" meant a more intimate study of the soul (psyche), and I immediately concluded that psychology as a discipline must be about this pursuit. This philosophical interest led me to enroll in my first introductory psychology course. Our text for this course was the first edition of Ernest Hilgaard's *Introduction to Psychology*. My reasons for entering this course were anticipated in the introductory chapter of Hilgaard's book, where the discipline and its boundaries were discussed, and this introduction was to disabuse me of my original intention for enrolling in the course. I was to learn that, in the 20th century, people who called themselves psychologists were no longer interested in perennial philosophical questions about the human psyche or person. In fact, these philosophical questions were considered to be obscurantist and passé. Psychology was now the "scientific" study of human behavior. This definition of psychology by Hilgaard was by no means idiosyncratic to this introductory textbook. In looking over other introductory texts, one could easily see that this was normative to the discipline as it attempted to define itself and its ultimate concerns.

I would be inclined to say that most students who take a course in introductory psychology have initial motivations somewhat similar to my own. As an early adult, it makes sense to try to engage in a deeper reflection on motivations, one's own and other people's—in short, what makes people tick. The promissory note that I received in this introductory chapter was that this was to be achieved if I made myself more of a scientist and less of a philosopher. Twenty years later, I am professionally a psychologist and not a philosopher, a testimony to the "normative power" of my socialization into that discipline. Let me now briefly discuss how this socialization occurs in more general terms.

In one sense, it is a misnomer to call psychology an internally coherent discipline. The relationships between clinical, personality, learning, and physiological psychology are tenuous at best. These topic areas appear side by side with one another in introductory psychology textbooks, and I would say that the closest relationship between these areas of concern is their proximity to each other (spatially, not conceptually) in the text. With some exceptions, most chapters within introductory texts can be reshuffled without changing the text appreciably. In a very important sense, these chapters appearing side by side are bureaucratic arrangements, that is, they are separate compartments with an interior life of their own existing side by side with other compartments having their own interior life. In giving this interpretation of a psychology text, I have also partly given a fairly good description of a psychology department. The bureaucratic arrangement of a text holds together more easily than the departmental arrangement because the text is not a dynamic, living thing. A department must present a coherent program of teaching and research, and this involves departmental meetings that bring these elements or compartments into the same room with one another. When this happens, conflicts arise and we hear of the perennial political struggles that are, in essence, conflicts in ideologies arising from the need to set priorities for a department. In many instances, the only integration (i.e., integrity) of a particular department's program is seen in the departmental catalog.

Yet all the personnel involved in this undertaking are willing to call themselves psychologists and would probably agree that their ideological enemy is a psychologist also. Despite their differences, they seem to share a common history of socialization that they are prepared to pass on to a new generation of students. I would like to hazard a guess as to some of the features shared in common by psychologists of different persuasions that give them a "professional identity" of sufficient integrity to be passed on to new generations. First, in the division of labor of the professions, psychologists have, relatively speaking, an interest in molecular versus molar events. Even though psychologists use group data, they are not, in contradistinction to sociologists, interested in the construct of group in molar terms. The typical commonsense way of saying this is that psychologists are interested in individuals while sociologists and anthropologists are interested in societies. The assumption here is that the different disciplines can do justice to these constructs separately. Second, psychologists are scientists and not philosophers. As the norms for psychology as a science have developed in this century, the science was to be a natural science with an emphasis on quantification. As opposed to philosophers who "armchair" it, psychologists as scientists have a desire to generate data and facts that are independent of their own cogitations. Although clinicians, experimental psychologists, and others will argue about what constitutes data or facts, they are all in agreement that data are important. Third, it seems to be understood that the single most important method of gathering data or facts is via the "experimental" method. Correlation studies come second and life-history and other types of interview data are much further down the line. There is no question that many psychologists, now and in the past, would disagree with these priorities for making observations. What they would not disagree to is that these priorities are generally normative for mainstream psychology. They would probably deem their own disagreement as countercultural to the mainstream. Fourth, psychology as a science is a theoretical (epistemological) and not a normative (ethical) science. In relation to "facts" and "theories" generated by psychologists, these are to be considered free from ethical norms (i.e., value-neutral). As a science, psychology is amoral and apolitical. This type of socialization allows a psychologist such as Jensen to regard his position on "genetics and racial equality" as a value-neutral and apolitical stance.

The book that you are about to read departs in substantial ways from these assumptions. If the set of assumptions stated above can be considered "normative" for the discipline without appearing to be strident, this book, it is hoped, offers a countercultural socialization to the reader. First, this book will attempt to hold the individual and society (i.e., molecular-molar) in dynamic tension. I assume that the person and society are dialectically related to one another, and that any attempt to collapse this tension would end up concealing more than it revealed. The separation of these constructs follows from the specialization and isolation of method within disciplines and subdisciplines. Thus, in order to achieve clarity and explanatory power, it was necessary to operate within a continually decreasing "circle of interpretation." The historical would be separated from the political, the psychological from the sociological, and so on, to the point where some would say that there was a crisis of imagination produced by the myopia of specialization. Ignoring, for the moment, the possibility that methodology in the social sciences replicates the social nature of production in the marketplace, it seems appropriate here to focus on this vacuum in imagination and how it affects inquiry. Here I would like to quote at some length from C. Wright Mills's *Sociological Imagination* (1959). Speaking of the general populace, Mills says:

> It is not only information that they need—in this age of fact, information often dominates their attention and overwhelms their capacities to assimilate it. It is not only the skills of reason that they need—although their struggles to acquire these often exhaust their limited moral energy.
>
> What they need, and what they feel they need, is a quality of mind that will help them to use information and to develop reason in order to achieve lucid summations of what is going on in the world and of what may be happening within themselves. (p. 5)

Second, in a broader interdisciplinary effort with an emancipatory intent, one would hope that psychology would provide a critical interpretation of what is happening within people in the context of wider social structures. Our assumption clearly is that the personal world exists, is influenced by, and influences the wider social context and cannot be considered apart from it. It is therefore necessary to be dialectical in order to pursue the approach I am advocating. Mills refers to it as a "sociological imagination," and for all intents and purposes we may consider psychology as critical interpretation as the development of a "psychological imagination." As Mills puts it:

> For that imagination is the capacity to shift from one perspective to another—from the political to the psychological; from examination of a single family to comparative assessment of the national budgets of the world; from the theological school to the military; from the considerations of an oil industry to studies of contemporary poetry. It is the capacity to range from the most impersonal and remote transformations to the most intimate features of the human self—and see the relations between the two. Back of its use there is always the urge to know the social and historical meaning of the individual in the society and in the period in which he has his quality and his being. (1959, p. 7)

In this attempt, we are not trying to make psychologists into philosophers or sociologists; rather, we are encouraging a certain philosophical reflection as an essential component of a psychologist's professional identity.

Third, this work will systematically attempt to develop a rationale for other modes of inquiry that are neither experimental nor correlational. We are not saying that the latter do not have merit. We are challenging their primacy as the "to be desired" mode of inquiry for theory and research in psychology. It is hoped that by the end of this book, alternative methods such as "life histories" and ethnographies will be seen as intellectually viable.

Finally, as a form of inquiry, it is hoped that psychology will be seen as a normative science. In the tradition of Emile Durkheim, we see psychology as an ethical enterprise. The position developed here is that the issue of "value neutrality" is an impossible, unrealistic, and—in the end—undesirable ideal for psychologists to be pursuing. We therefore have to tackle all the issues that follow from viewing the pursuits of psychologists as ethical enterprises. In order to do this, there must be a radical shift in ground from a theory (episteme) to practice (normative) working dichotomy to that of *praxis orientation*. A praxis orientation starts from the point of view of *human action* and formulates a reflection on that action (i.e., theory). This point of view is not a reverse dichotomy (i.e., practice to theory). Praxis assumes that human action is *intentional* in nature and, therefore, that reflection is embedded in human action where human actions are concerned.

Finally, the radical orientation being proposed shifts the interest nature of the definitional structure from one of control to an emancipatory interest in human freedom. This emancipatory concern needs unpacking, which is the reason for this book. At this point, one can say that an emancipatory interest in human freedom rests on the assumption that part of all human action is *creative* and on the assumption that humans create their world while, at the same time, being determined by it. The human intentions that can motivate human actions are history making. We call the latter function "culture." Up to this point, the psychological enterprise has been, we might say, "uncultured." This book intends to add a cultural dimension as the essential location where human freedom operates.

ACKNOWLEDGMENTS

There are many persons and groups that have influenced my thinking and thus helped in the writing of this book.

I would like to acknowledge a collective dealing with psychology and political economy that I sat with in New York City for several years; its members included Leon Rappaport, John Broughton, Joel Kovel, Adrienne Harris, Ricardo Zuniga, Howard Gruber, Harry Garfinkle, and Arnold Kaufman.

Also, the Critical Pedagogy and Cultural Studies Working Group at the Ontario Institute for Studies in Education in Toronto: Paul Olsen, David Livingstone, Jack Quarter, Mal Levin, Madan Handa, Gord West, and Roger Simon.

To friends at St. Michael's College: Bernard Black for gracious use of the library during sabbatical; also, Gregory Baum and Lee Cormie, theologians at St. Michael's, who encouraged my efforts at important stages of this work.

To friends among my colleagues: Susan Eadie, Dian Marino, Claudio Duran, Paul Callaghan, Howard Richards, Dieter Misgeld, Pat Lee, Jeri Wine, Cliff Christensen, John Weiser, and Jim Fowler.

Specific thanks to my PATH in Psychology editors, John Broughton and Robert Rieber. To Morry Ulrich and Donaleen Hawes for judicious proofing of the final manuscript. Enormous appreciation to my secretary, Diana Postlethwaite.

Finally, to my wife, Pat, the person who helps me nurture a sense of justice with our children and to whom this book is dedicated.

I depart from the usual "I do not hold them responsible for my ideas." I hold them all responsible in small and big ways.

CONTENTS

CHAPTER 1

Metaphors of Understanding: A Critical Introduction.   1

Metaphors of Understanding   1
The Mechanical Metaphor.   3
The Organic Metaphor.   8
The Personal Metaphor   17
Prospectus   30

CHAPTER 2

Human Expression: A Relational Act   33

Prologue.   33
Expression   33
Loose Ends: Habits and the Unconscious   47
Summary.   48

CHAPTER 3

The Personal World: A Relational Event, a Cultural Reality   51

Individualism: An Impediment.   51
The Personal World as a Relational Totality.   55
The Personal World as a Metaphor of Communication.   57
Integrative Components of a Personal World   60
Summary.   73

CHAPTER 4

Personal World, Cultural Forms, and the Structures of Class, Gender,
Ethnicity, and Age   75

Introduction.. 75
Power, Agency, and Domination............................ 75
Class as a Structure of Domination......................... 77
Class, Cultural Form, and Ethnicity........................ 88
Gender and Domination................................... 99
Age and Class Dynamics 106
Conclusion 108

CHAPTER 5

Interpretation... 111

The Personal World as a Text 111
The Problem of Mediation and the Paradoxes of Psychological
Interpretation... 113
The Hermeneutic or Interpretive Spiral 116

CHAPTER 6

Critical Interpretation..................................... 123

Emancipatory Psychology and Its Relation to a Critical Theory of
Society... 125
Psychology in the Social Order 130
Cause and Intention 141

CHAPTER 7

Adequacy of Interpretation 143

The Problem of Multiple Horizons 144
Study 1: The Rules of Disorder............................ 150
Study 2: Learning to Labour 156
Adequacy of the Accounts: A Comparison 166
Method Revisited... 173

Appendix.. 177

Bibliography... 181

Index... 189

METAPHORS OF UNDERSTANDING
A Critical Introduction

> The history of thought is the history of its models. Classical mechanics, the organism, natural selection, the atomic nucleus or electronics field, the computer: such are some of the objects or systems which first used to organize our understanding of the natural world and have been called upon to illuminate human reality.
>
> —Jameson, 1972, p. v

The title of this book is A *Critical Psychology: Interpretation of the Personal World*. This chapter attempts to develop an outline and elucidation of the parameters of that definition. Part of the definition of something is to indicate what it is not or what it should not be. Thus, there will be some critical discussion of what one might term "mainstream psychology." I commence this treatment under the assumption that psychological perspectives (e.g., behaviorism, structuralism) are conveniently thought of as metaphors. Ultimately, my intent is to develop a definition of psychology based on a personal metaphor (cf. MacMurray, 1957, 1961). At the level of commonsense language, it seems likely that no psychologist would deny that he or she is attempting to understand people (i.e., persons). However, much of what constitutes psychological theory and research is not couched at the level of ordinary language. In fact, most of the language of psychological theory quickly embraces a formalization of symbols that is oriented away from ordinary language. As we shall point out, formal language is, in part, an attempt to go beyond the vicissitudes of history, which are clearly embedded in our ordinary use of language. In all this, it should be noted that, whatever way we turn, we are embedded in some kind of linguistic system, a "prison house of language" (Jameson, 1972).

METAPHORS OF UNDERSTANDING

The world itself and people within it are relational. By this we mean that to know or identify persons, things, or events, we have to identify or contrast them with other persons, things, or events. In order to come to know and understand something, we usually attempt to identify and contrast it with something that we already find familiar. Piaget calls this "assimilation," which means that we attempt to come to an understanding of the unfamiliar by use of the familiar. This process can be identified as metaphorical; that is, we use metaphors in order to say something that we know little about (Te Selle, 1975). The dictionary defines a metaphor as a figure of speech denoting a word phrase, usually one kind of object or idea in place of another, in order to suggest a likeness or analogy between them. A metaphor is, at the very least, an interaction between symbols. Black sees this interaction as a relational event where the conventional wisdom associated with one context serves as the screen or grid through which we see the other context:

> Suppose I look at the night sky through a piece of heavily smoked glass on which certain lines have been left clear. Then I shall see only the stars that can be made to be on the lines previously prepared upon the screen and the stars I do see will be seen as organized by the screen's structure. We can think of a metaphor as such a screen and the system of associated commonplaces of the focal word as the network of lines upon the screen. We can say that the principal subject is "seen through the metaphor expression" or if we prefer that the principal subject is projected upon "the field of the subsidiary subject." (Black, 1962, p. 41)

Another term that is a species of metaphor is *paradigm*, introduced by Kuhn in his *Structure of Scientific Revolutions* (Kuhn, 1962). I am reluctant to use this term in our present context since it tends to convey the feeling of a fully developed and conscious theoretical perspective. Though there are many attempts at systematic formalization of theory in psychology, none of these adds up to or commands the allegiance of the paradigm in the Kuhnian sense. We are, therefore, restricting ourselves to the use of the term *metaphor*. Our use of this term alerts the reader to the fact that we are using language as a "lens" for understanding. To conceive of psychological perspectives as metaphors for understanding means that the problems presented by language affect all theoretical perspectives of whatever variety. Therefore, all theoretical perspectives structure perception and practice as a prism filters light. As metaphor, theoretical language both reveals and conceals different aspects of human reality (Ricoeur, 1970). All language is embedded in certain tacit prerequisites that are often unknown to those utilizing it (Gadamer, 1975a; Polanyi, 1968). For example, language is embedded in sociohistorical realities and it reflects those realities (cf. Marx & Engels, 1960). When those sociohistorical realities are denied, the theory language is said to be "ideological," that is, it supresses some of its own intentions (Habermas, 1970, 1972, 1974; Marx & Engels, 1960). Therefore, in treating theories as metaphors, we are assuming that the prism of language is embedded in sociohistorical realities and not independent of them. Thus, behaviorism is not only a metaphorical language but also a social practice embedded in sociohistorical realities. The praxeology of a metaphor is often denied in the development of the metaphor. For example, psychometric psychologists have, in the past, denied that their perspective on the nature–nurture question was influenced by and influenced immigration policy (see Kamin, 1974). This example is used simply to alert the reader to the fact that all language, including theory language, is embedded in sociohistorical events. This, at any rate, is the perspective out of which I am operating. With this understood, I would now like to discuss psychological theory and practice under the aegis of three root metaphors: the mechanical, the organic, and the personal. This particular breakdown is suggested by John MacMurray (1957, 1961), and it must be understood at the outset that this particular breakdown is, to some extent, arbitrary in character. I choose this particular constellation in the belief that it will be helpful to a person trained as a psychologist. I am saying, in essence, that these three root metaphors conveniently encompass psychological theories that serve or have served as screening devices in our understanding of the human world.

The Mechanical Metaphor

Machines serving as analogues for the understanding of human behavior have probably always attracted human attention. The attraction became systematic when machines came to the forefront of consciousness at the time of the industrial revolution. During the 16th and 17th centuries we were confronted with a new view of the universe that is both mechanistic and materialistic. Mechanical materialism arose out of the developments achieved in the physical sciences, particularly physics. Mechanism, that is, seeing the physical world in terms of machine analogues, came to be seen as the most important explanatory language in the development of what are now the physical sciences. From the point of view of mechanistic materialism, the world consists of particles of matter in interaction. Each particle has its own existence and becomes a large totality by addition. The totality of interactions between particles forms the totality of everything that happens in the world, and these interactions are mechanical in nature; that is to say, they consist of external influences of one particle upon another. As such, mechanistic materialism saw the whole world as nothing but a complex piece of machinery, a mechanism (Cornforth, 1971). Although originally designed as a metaphor for reflection on phenomena in the physical world, mechanistic materialism historically, during the latter part of the 19th century, became a controlling image for reflection on social phenomena. In the 20th century, the discipline of psychology appropriated mechanism from the physical sciences and used it like a "borrowed tool box" (Hampden-Turner, 1971). The appropriation of "mechanical tools" opened the door to looking at human events as mechanisms not unlike complex machines. Hull (1943), the famous behaviorist theory builder, refers to humans as "complex robots."

Instrumental Rationality and Technical Control

The appropriation of the metaphorical language of mechanistic materialism involves a complex synthesis of several interrelated notions. For example, the pervasive *rationality* exemplified in mechanical metaphors is a combination of logic and empiricism (i.e., logical empiricism). Logical empiricism is a complex amalgam of the historical traditions of rationalism and empiricism with empiricism at the forefront; as we shall see later, structuralism reverses these priorities. Logical empiricism exists within a given societal matrix, so that any form of understanding of, and orientation to, a scientific and technical power needs to elucidate this reference to the investigator's membership in society (Gadamer, 1975a, 1975b; Habermas, 1970). The matrix or metascience of early American psychology was essentially logical empiricism. Radnitzki (1973) ventures that the typical human scientist of the Anglo-American world appears to be programmed by logical empiricism. The dominant image within this matrix places human "behavior" at the center, and it is studied in relation to its component parts (i.e., analysis). At the level of ideas, the philosophy of science concomitant with this image of the human is characterized as "reductionistic," methodological "individualism," and so on. Psychology in its essential defining characteristics becomes a behavioral science, typically behavioristic, physiological, or psychometric in nature (Radnitzki, 1973). Methodological individualism means that the ideal person is considered as a self-contained individualistic system rather than an interdependent system (Sampson, 1977). At the order of events, it may be contemplated that the methodological individualism reflects the historical individual of the surrounding culture (Sampson, 1977). Within the context of the American ethos at the beginning of this century, the budding and neophyte psychological establishment attempted to establish an experimental psychology (i.e., empirical) which would supply the fundamental laws governing all human activity, irrespective of context (Danziger, 1978). Psychology, either knowingly or unknowingly, was entrenched in the American socioeconomic order, and its most articulate spokespersons would express themselves as creating a science that would help us to understand the motives that control the actions of labor and capital. Watson, an early president of the American Psychological Association, suggested that psychology accept the premise that its discipline was to produce data in a practical way to be easily utilized by businesspersons (Ewen, 1976). American psychology was to be an *administrative* science: a technology to be willed by society's managers to direct the actions of those in their charge into desired channels (Danziger, 1978). Although there would be other reasons, it is not farfetched to conceive that for management reasons, psychology would take as its primary definition the prediction and control of behavior.

Analysis. The whole is the sum of its parts; that, in essence, is the principle of analysis. It is an epistemological principle assumed in many areas of inquiry.

It is contrasted with the synthetic principle that the whole is more than the sum of its parts. The *principle of analysis* is clearly accented when mechanical metaphors are used in inquiry. One can easily assent to the idea that the construction of a machine is based on the sum of its parts and that the finished product can be decomposed back into its component parts. As I will discuss shortly, the mode of observation called the *experiment* represents a decomposition process with the intent of finding explanations of and for events. When the principle of analysis is accented, it is assumed that manipulation, prediction, and control of supposedly decomposed conditions are explanations of events. *Explanation* of a phenomenon involves an attempt to manipulate, control, and predict antecedent–consequent conditions with a view to defining linear causality. Causality is a relationship between antecedent conditions and consequences. Manipulation and control assume the ability to control some conditions in nature, to vary these conditions while holding other conditions constant, and then to observe differential consequences in this controlled setting. The stress is on the ability to control the consequences through the systematic variation of antecedent conditions. If antecedent conditions are related to predictable consequences (e.g., if a, then b); a causal relationships between these events is assumed. When these component parts are human events, complete psychological determinism is assumed. *Prediction* follows from manipulation and control of conditions. Prediction assumes determinism, and the aim of prediction is to achieve *causal explanations* of a set of events or phenomena. Explanation is, therefore, said to be a result of the predictive validity of the manipulation and control of the experimenter. A good explanation of an event should have a high predictive component.

It is believed by those who foster it that the psychological experiment achieves some kind of disinterested rationality and, as a technique, is independent of its historical surroundings. My own contention is that the experiment is a reflection of a culture whose reference point is a rationality based on technical control (cf. Habermas, 1970, 1972). In other words, the knowledge is based on a particular *interest* reflected in modern scientific society, and this knowledge-constitutive interest (i.e., technical–control rationality) shapes what counts as knowledge. Thus, technical interest aids in the formulation of the categories relevant to what one takes to be knowledge claims. A rationality formulated as technique is, of necessity, a particular type of rationality governed by technical rules (e.g., the experiment proper).[1] Technical control rationality is connected to Habermas's (1972) description of conventional empirical–analytic sciences. Behaviorism is one psychological variant of what Habermas calls "conventional

[1] It should be clearly understood that the psychological experiment is only one instance of a technical control rationality. Test construction and the statistical techniques used in assessment are another specific example. The experimental method, for some historical reasons, appears to have eclipsed all other techniques as the most substantial technique for the acquisition of knowledge.

empirical–analytical science." It is a family of theoretical and research thrusts that includes several different points of view of how one arrives at true statements of facts about events (i.e., explanation). Hull's (1943) behavior theory was a rather elaborate molecular theoretical system (i.e., hypothetical–deductive) combined with a rigorous tying together of the theoretical system and the empirical data. Hull's system is a combination of deductive theoretical statements (i.e., rationalism) interlocked and intertwined with empirical statements (i.e., empiricism). This is logical empiricism in its clearest formulation. Skinner (1950) eschewed the elaborate theoretical formulations of Hull and others and preferred formulating his behaviorism with a rigorous inductive empiricism. His article entitled "Are Theories of Learning Necessary?" allows one to label him a radical empiricist. We are not using the term *radical* here in a political sense. Skinner is radical because his conception of learning rejected the elaborate theoretical construction characteristic of Hull and Tolman (e.g., Hull's elaborate hypothetico-deductivism).

At the level of practice, all behaviorists of whatever theoretical persuasion agree that the experiment is the preferred mode of verification. Even though there are important practical differences between the ways in which classical and operant conditioning are researched through experiment, all the various experimental formats are alike in their preoccupation with control over what is being studied. All experiments in their pure state are mechanical procedures for the purpose of explanation.

Objective Knowledge. The experiment as a sound practice unreflectively collapses an important epistemological dimension, which I think is important not to overlook. The dimension I am referring to is the dialectic of subjective and objective meaning. The experiment, in practice, is a denial of the importance of subjective meaning. If the mind or consciousness is simply a mechanism in nature, subjectivity assumes minor importance within this perspective or is totally ignored. This is behavioral psychology's classical attitude toward consciousness and subjectivity; for the sake of truth, it should be noted that this attitude has eroded in more recent years.

Fact–Value Dimension. Another dimension that collapsed within the experimental praxis is the normative element (i.e., values). The assumption is that the experiment is an enterprise that is value-neutral. As a methodological ideal borrowed from the natural sciences, it is said to be a method of establishing "true facts." This type of restriction allows some real ethical problems to go unacknowledged (e.g., the Milgram experiment).

The principle of analysis assumed in the experiment systematizes an ignorance toward larger "totalities" in which the experiment is contextualized (i.e., institutional–power relationships). This model has recently come under serious challenge as a result of its epistemological horizon and its value-neutral stance (Gadlin & Ingle, 1975). Experimental self-reflection is a one-way mirror that reflects the technical control rationality of the experimenter, while ignoring the subject's or animal's point of view. Particularly where human subjects are involved, there is a systematic sundering of communication between the experimenter and the subjects. The use of the term *subject* is quite telling in this respect.

It is now becoming apparent that the "detached observer" role of the social scientist possibly masks a situation of domination and control of one group of persons over others. Essentially, the experimental method as a mode of reflection illustrates the condition where there is a minimum of reciprocity between those being studied and the person (i.e., social scientist) who is carrying out the inquiry. As a reflective process, it is seen by Gadlin and Ingle (1975) as a one-way mirror. The mode of reflection generates a process that bends the social reality of communication in one direction (i.e., unilateral), the relationship of communication being one of monologue rather than dialogue. Thus, in the attempt to adopt the viewpoint of a *detached observer*, the experimenter fragments the relationship further by not allowing himself or herself to be influenced by the social reality of those being studied. Upon challenge, a hidden dimension of control is revealed which, under normal circumstances, is taken for granted. The viewpoint of the detached observer masks a power relationship that appears to be hidden from the researcher. The experimenter (i.e., social scientist) is unaware of the fact that he or she is in a power relationship with those who are being studied. From the point of view that we will be developing, the absence of reciprocal communication represents a type of oppression. The communicative position of the experimenter is like that of the Producer in Pirandello's (1952) *Six Characters in Search of an Author*:

> Producer: That's the idea! And if you want my opinion you ought to be grateful for all the attention they are paying you.

> Son: Indeed! Thank you! But hasn't it dawned on you yet that you aren't going to be able to stage this play? Not even the thinnest vestige of us is to be found in you—. And all the time your actors are studying us from the outside. Do you think it's possible for us to live confronted by a mirror which, not merely content to freeze us in that particular picture which is fixing our expression, has to draw an image back to us which we can no longer recognize? Our own features, yes—-. But twisted in a horrible grimace. (Act 3, p. 274)

A reflexive psychology reveals the value arrogance involved in experimental communication. When the communicative values implied in this perspective are shown to be reflecting wider sociopolitical realities of a repressive nature, they are vigorously denied (see Zuniga, 1975). Denial does not, however, detract from the reality that this type of social science praxis masks political intentions. The technical control rationality is clearly sympathetic to identifiable political intentions (Habermas, 1970). Thus value neutrality cannot, in the end, be retained as a viable concept for inquiry.

Computer Simulation

That psychological theory is culture-contextual is evident with the development of computer systems. With a shift toward complex machine technologies, as witnessed by computer simulation, we see a shift in theory and practice where mechanical metaphors are concerned. The computer is seen to hold out the possibility that the simulation of complex mental processes is now possible (e.g., Newell & Simon, 1961). Neisser (1967) exemplifies the machine-metaphorical use of the computer as a guide to human understanding, as follows:

> The task of the psychologist trying to understand human cognition is analogous to that of a man trying to discover how a computer is programmed. In particular, if the program seems to store and reuse information, he would like to know by what "routines" or "procedures" that it is done. . . . By the same token, it would not help the psychologist to know that memory is carried by RNA as opposed to some other medium. He wants to understand its utilization, not its incarnation. (p. 6)

Recently Neisser (1976) has become disenchanted with his earlier views, and he probably reflects a growing skepticism of the long-term possibilities of the computer analogue as an adequate reflection of human capabilities. At the same time, it should be acknowledged that there is still substantial attraction for computer simulation research, and this will continue to increase with the new communications technologies. I am not here embarked on an elaborate analysis or critique of computer simulation and its currency in contemporary psychological theory and research. It is briefly discussed here to show how mechanical metaphors are transformed by the development of machine technology.

It is, however, important to question the extensive use of the machine metaphor for the full range and diversity of human functioning. As a metaphorical lens or prism, I am cautiously observant of developments in this area.

THE ORGANIC METAPHOR

If mechanical metaphors are premised on physical systems, the metaphoric prism of organic metaphors is premised on biological systems (i.e., life sciences). The term *organic* refers to the processes or products of life—in human beings, animals, or plants (Williams, 1976). Concurrent with the development of the life sciences, there is the development of the distinction between organic and mechanical systems. The fundamental difference between the mechanical and the organic metaphor is essentially that an *organism* is a living, adaptive unity that *transforms* itself while relating dynamically to its environment. Probably the most important single theory that exemplifies the organic metaphor in contemporary times is Darwin's theory of evolution.

By the 20th century, a broad distinction between biological evolution and social evolution had been articulated in diverse areas of the budding social sciences. The early 1900s witnessed a marked interest in social evolutionary theory in sociology and anthropology (Campbell, 1975). This interest has waxed and waned in 20th-century psychology, but it is on the ascendant again with a new social science concern labeled sociobiology (Wilson, 1975).

For the purposes of exposition, we will restrict our discussion of the organic metaphor to two major contemporary movements of importance: structuralism and functionalism as specific exemplars.[2]

Structuralism[3]

In the late 1950s and early 1960s there appeared a rather interesting transitional period in American psychology. There was a crack in the wall of the hegemony of logical empiricism. Behaviorism was becoming much more complicated and there was a movement toward a behavior theory that could deal with the complexities of the human organism. For the first time, we see an attempt to address complex linguistic processes by incorporating the term *meaning* (Osgood, Suci, & Tannenbaum, 1957). The logical empiricism formerly presented by behaviorism was being stretched to its limits. New theoretical shifts were taking place at this time, which would eventually lead to the decade of structuralism in American psychology. Structuralism, as specifically presented to the psychological scene, was to be an essential challenge to the prevalent mechanistic theories. When the Swiss genetic epistemologist Piaget entered the American psychology scene in the 1960s, there had already been a movement toward centralist or mediated behaviorist formulations. The psycholinguist Charles Osgood was suggesting complex mediational processes to explain linguistic meaning. Behaviorism at this time was attempting to come to terms with complex mental processes, and we also saw information processing models spinning off from computer simulation research (Hilgard & Bower, 1966). Behaviorists (e.g.,

[2]Our treatment here of just two examples of organic metaphors is in no sense exhaustive. Our selection is made simply for illustrative purposes. For example, under organic metaphors one could include humanistic psychologists such as Maslow and Perls. The distinction between structualism and functionalism is fraught with ambiguity. Piaget is called a structuralist, but his theory is also embedded in functional definitions (e.g., intelligence). Parsons is considered a functionalist, yet his perspective is informed by structural dimensions. Our distinction here between structuralism and functionalism is simply on the basis of accent. Some theories emphasize and articulate structural dimensions over function (e.g., Piaget, Chomsky). Thus our appellation of the term *structural* does not preclude functions and vice versa.

[3]To give structuralism the appellation of an organic metaphor demands some reservations on my part. I am aware that other authors might not follow me on this point (see Jameson, 1972). The truth is that I am simplifying issues for the purpose of illustration. Specifically, I am accenting those aspects of structural theories which clearly exemplify aspects of what I am calling "organic metaphors."

Berlyne, 1966; Gagné, 1968) and computer-simulation theorists attempted a rapprochement with Piaget's and Chomsky's structuralism. For the most part, these attempted marriages had ill-fated results (e.g., Newell & Simon, 1961; Hunt, 1961). The reason is that, on a fundamental level, structuralism deviates radically from mechanistic theories, even in its complex guises, because there is a synthetic rather than an analytic conception of knowledge. By *synthetic* we mean a preoccupation with structural totalities (i.e., the whole is more than the sum of its parts).

Structuralism is one of several attempts to go beyond some of the dilemmas presented by the methods of logical empiricism (Sullivan, 1980a, 1981; Unger, 1976). A structural conception assumes that a structure (in whatever form it takes) is a *totality*; that is, the whole (i.e., structure) is more than the sum of its parts. In contrast to mechanistic explanation, which is fundamentally an analytic process, all forms of structuralism are based on a principle of synthesis (Unger, 1975). The structure, called "stages" in the work of Piaget, is a holistic entity in which the whole is more than the sum of its parts (Sullivan, 1966, 1980a). As already mentioned, structure as a synthetic principle is diametrically opposite to analysis, as exemplified by even the most complex behavior theories (e.g., Berlyne, 1966). A structure is a holistic entity characterized by *internal* dependencies. This is explicated by Piaget, for example, in his definition of a structure as a system of transformations which, as system, implies a lawfulness of organization independent of the elements that compose it (Dagenais, 1972). This system is characterized as a totality. In other words, whatever the composing elements in the system, they are subordinated to the laws defining the system as a system. In addition, a system is characterized by multiple transformations that are interdependent with each other and with the totality, that is, they are dependent on the structure itself. Finally, a structure is self-regulating and tends toward the conservation and enhancement of the system itself (Dagenais, 1972). Similarly, in linguistics, Chomsky and his followers deny that linguistics can be built on the analysis of elementary particles of meaning (e.g., analytic assumptions). Transformational grammarians share the fact that the speakers of every language are able to construct an infinite number of formally correct and meaningful sentences. Linguistic structuralism has, therefore, been highly critical of behaviorist interpretations of how language is acquired. Consider Chomsky's observation:

> I believe that the study of human psychology has been diverted into side channels by unwillingness to pass the problem of how experience is related to knowledge and belief, a problem which of course presupposes a logically though not necessarily temporally prior investigation of the structure of systems of knowledge and belief. No matter how successfully the study of stimulus-response connections, habit structures, and so on is pursued, it will always fail to touch these central questions. The systems of knowledge and belief that underlie normal human behavior simply cannot be described in terms of networks of association, fabrics of dispositions to respond, habit structures, and the like. At least this seems to be true in the case of language and other known examples of human "cognitive processes." (Chomsky, 1971, pp. 47–48)

It is safe to say that the structuralism of Piaget and Chomsky deviates substantially from all types of neobehaviorism because of its interest in synthetic totalities (e.g., generative grammars, stages of cognitive and moral development, etc.).

Just why structuralism should gain such power is a matter for some complex historical speculation; but it is clear that it is not just the result of the explanatory power of the theories proposed. It was surprising that within a few short years, Piaget's study of cognitive development (in its nascent stages), Chomsky's conceptions about language development, and Kohlberg's articulation of moral development were to become the popular psychological metaphors on the American psychological scene. Wilden (1975) has characterized structuralism, among several psychological currents, as a veritable epistemological shift in the way of conceiving living and social systems. He links this to a progressive transformation of the large socioeconomic structure surrounding psychological speculation:

> The increasing size and complexity of business corporations, as well as their organization in tiers, trees, pyramids, and more intricate topological structures would naturally lead to the systematization of a body of theory designed to explain and control organization. Similarly, one has only to consider the relatively sudden discovery by the industrialized nations that the pursuit of individual economic independence leads ultimately to increasing degrees of collective inter-dependence to realize that we would necessarily come to live in an era when some form of systems ecology would be applied to any and all complexities of system-environment relations. The shift is both a mirror of contemporary socioeconomic reality and an illumination of it. (p. 91)

There has gradually been a shift away from mechanical metaphors that talked about biological and human relations in terms of aggregates, entities, atoms, individuals, closed systems, linear causality, and so on (Wilden, 1975). The new vocabulary employs such terminology as *wholes, interdependent structures, open systems, feedback, information,* and *communication* (Wilden, 1975). Although this shift in epistemology is a contemporary phenomenon, it has historical roots that go back several centuries (Chomsky, 1968; Taylor, 1971). Even as early as Descartes, there was a belief that mechanical metaphors in their nascent state could not explain the complexities of the mind in terms of quantity—that, rather, a qualitative dimension was necessary in any adequate form of explanation. It has taken well over two centuries to replace a traditional and quantitative epistemology with a multidimensional viewpoint of levels and complexity that necessarily involves a comprehension of qualitative considerations as a prerequisite for any quantitative analysis (Wilden, 1975). Dagenais (1972) notes that structuralism rose out of the amalgam of psychology, anthropology, and sociology, passed through the science of linguistics, and has passed back into psychology, psychoanalysis, anthropology, and sociology.

To summarize, structuralism in its psychological and linguistic manifestations represents a substantial change in orientation from classical logical empiricism. In contrast to the methods of behavioral psychology, for example, it does not attempt to order phenomena at the level of empirical events with the objective of demonstrating causality between events. Instead, the method of structuralism seeks, through a process of interpretive abstraction, to tease out the underlying structure of phenomena. Its controlling metaphor is *organism* rather than *physical systems*. It therefore draws its interpretive language from *biology* rather than *physics*, freely utilizing global evolutionary concepts as part of its interpretive system.[4] Structuralism deviates markedly from mechanism at the level of epistemology. The "structures" that are articulated by structuralists are "wholes" or "totalities." A structure is, therefore, more than the sum of its parts. Historically, it is an offshoot of Cartesian rationalism and Kantian *a priorism*. By postulating the existence of *a priori* categories (i.e., structures), which are *active* organizing principles for knowledge acquisition and so on, this method at least concedes something to the notion that meaning is constituted by actors. This is achieved, unfortunately, by severing the link between the ideas of subjectivity and of consciousness (Sullivan, 1980a). In fact, around the issue of the importance of the subjective intentions of actors (i.e., agents), structuralists are anti-phenomenological (Petit, 1975; Piaget, 1971). Structuralists, by and large, reject the standpoint of a psychology based on *intention* of *agents*. The articulation of the structures is from the viewpoint of the social scientist articulating the structure and is not the point of view of those who are studied. As we shall see shortly, this is one of the dimensions in which structuralism differs from a personal metaphor that incorporates the conscious intentions of agents as part of its interpretative practice.

Functionalism

A functionalist position, as I understand it in a psychological context, considers mental processes of sense perception, emotion, volition, and thought as functions of the biological organism in its adaptational effort to influence and to control its environment. Specifically, where psychology has been concerned, this orientation has been involved with the articulation of "traits" that are considered to be adaptive or maladaptive for individual and societal survival. Human traits *per se*, within this perspective, are considered to be adaptive organs.

Intelligence Testing. Although many human traits have been of interest to psychological research (e.g., intelligence, altruism, competitiveness, aggression, etc.), by far the most important trait to have been studied is that of intelligence. A broad definition of *intelligence* is "a trait of cognitive judgment that is present in degrees in individuals and groups and is said to have adaptive significance."

---

[4]It must be pointed out that behaviorists have also been introduced to evolutionary theories. Nevertheless, in contrast to organic metaphors, the use of evolutionary concepts is mostly tacit rather than explicit.

The emphasis on degrees expresses the fact that intelligence may be considered a quantitative trait that can be measured and assessed. The testing movement also assumed that individuals and groups had more or less of that trait. Proponents of the testing movement and the testers proper assumed themselves to be in a value-neutral position when making assessments and comparisons between different groups of people. It is within the context of this movement that the nature–nurture controversy was carried out in the past and is still active in the present (e.g., Jensen's 1969 position). The assessment of traits was accomplished through a set of standardized tests based on large-sample statistical comparisons. Because of the pretense of value neutrality, the testers could legitimate broad social policy questions from a position of dispassionate interest. IQ testers assumed themselves to be generating "objective evidence about different population norms." Kamin (1974) has demonstrated the bias of the testing movement and its essentially conservative political functions. It helped to justify essentially repressive policies about different groups and races. Kamin's (1974) historical treatment of the early tester reveals them as essentially racist and sexist by today's standards (e.g., Jews were considered by some testers as an inferior race, thus justifying specific immigration policies).

The assumptions of the testing movement were that certain traits, and their presence or absence in different populations, were socially adaptive to this movement, essentially wedding a functionalist biological perspective to important social questions (e.g., immigration policy, school placement, work placement, etc.). What is important to note here is the position that the testers took vis-à-vis the individuals or groups they were testing. The "objective test" was designed by the tester and usually defined what was adaptive or maladaptive in advance of any conception of adaptation that those tested might have had. The assumptions of what was adaptive or maladaptive were defined in an ethnocentric manner unknown to the testers, since they assumed a certain world view as superior (i.e., white Anglo-Saxon). This nonreflexive approach is still very much apparent in current objective testing. The ethnocentrism of the testing movement has also seriously underestimated the diversity of the linguistic world of cultural groups.

Another variant of psychological functionalism is the whole area of trait differences that has developed around differences between men and women. Of recent popularity is the work of Horner (1972) on the study of women, which explores what she postulates as a motive to "avoid success." Thus, achievement in a social setting is reduced to a social/psychological trait, presumably, more ascendant in women and accounting for their reputed lack of success in achievement settings. The argument for the "fear of success" motive follows a form common to biological functional metaphors. Eichler (1979) captures this form succinctly:

> One biological difference between the sexes, presumed or real, is seized upon and
> universally applied to all human interactions, be it a metabolic difference between the

> sexes, a hormonal difference, or a general inferiority of development in women. (Eichler, 1979, p. 336)

In continuing this position, Eichler (1979) maintains that it is not possible, in principle, to explain social relations in raw biological terms. Conceding that biological factors can play a role in determining social relations (e.g., longevity, infant mortality rates, etc), she nevertheless stresses that these biological conditions in humans are mediated through culture. This is the position I will develop in further detail when gender is discussed more thoroughly in a later chapter.

Sociobiology. The most recent discussion of an interdisciplinary inquiry called "sociobiology" has redirected the focus on cognitive traits such as intelligence to those personality traits that enhance the cohesion of human groups (e.g., altruism) or detract from group cohesiveness (e.g., aggression, competitiveness). This emphasis is partly in response to a perceived crisis in cultural solidity that has accrued historically from the traits encouraging an unrestricted and overriding individualism (Campbell, 1975). Here it is necessary to give to the reader a provisional definition of *sociobiology* so that there is a context for our discussion of it as an organic metaphor. Caplan (1978) defines it as follows:

> Sociobiology is the latest and most strident of a series of efforts in the biological sciences to direct scientific and humanistic attention toward the question of what is, fundamentally, the nature of human nature. Sociobiologists believe that evolutionary inquiry into the origins and developments of social behavior in organisms will shed great light on traditional philosophical puzzles concerning the ability of humans to be both influenced by and freed from the limits set by their biological constitutions. (p. 2)

The recent publication of Wilson's (1975) Pulitzer Prize-winning book *Sociobiology* represents an attempt to establish the social sciences as a branch of evolutionary biology. The potential relevance of this field for psychologists can be seen from a recent presidential address given to the American Psychological Association (Campbell, 1975). In this address, Campbell encourages every psychologist to own this volume, whether she or he agrees with it or not. Campbell (1975) takes from sociobiology the broad evolutionary framework that it offers and the desire to interpret human social processes within a biological rather than mechanical framework. Critics of sociobiology see important and vexing questions arising when biological, organic, evolutionary metaphors move into the areas that distinctly characterize the human (i.e., rationality, morality, culture, mentality, linguistic ability, and intentionality). Caplan (1978) asks whether it is possible to explain priorities such as these by means of purely biological parameters. Campbell, responding in the affirmative, encourages psychologists to pursue the broad outlines of a sociobiology in their own specific work. Campbell (1975) ventures a dialectical treatment of biological and social evolutionary perspectives and tries to assess the significance of social norms and institutions for individuals, with special emphasis on altruistic traits. He assumes, as does Wilson, that our behavior is a product of evolutionary adaptation and is probably designed to maintain a stabilizing effect in a society. He also assumes, with Wilson, that humans have achieved a very high level of cooperation with little sacrifice of personal survival. Cooperation, it is assumed, may well confer a "selective advantage." Two conclusions about humans stem from the "genetics of altruism":

> (1) Human Urban social complexity has been made possible by social evolution rather than biological evolution. (2) This social evolution has had to counter individual selfish tendencies which biological evolution has continued to select as a result of genetic competition among cooperators. (Campbell, 1975, p. 115)

Campbell (1975) assumes that humans have selfish tendencies because biological evolution would select these as a result of interindividual genetic competition. Social order is achieved through norms for altruism derived from social evolutionary processes, as contrasted to biological evolution. These altruistic norms function to override or limit genetic selfishness.

I am not here developing an extensive critique of sociobiology, since it would divert us from our primary purpose. I mention Campbell's suggestions because he is attempting to show the relevance of a sociobiological perspective to psychologists. I do, however, question the ultimate utility of the superordinate use of organic metaphors, such as are used in the superorganic perspectives suggested by sociobiological orientations. My eventual discussion of a personal metaphor suggests that the reductionism implied by mechanical and organic metaphors is ultimately shortsighted for an integrated conception of the human. My own contention in developing a metaphor of the personal is the assumption that human language is an essential and irreducible constitutive element in our understanding of humans. In criticizing sociobiology, Sahlens (1976) brings out the point I am making here:

> If we were to disregard language, culture would differ from animal tradition only in degree. But precisely because [of] this "involvement with language"—a phrase hardly befitting serious scientific discourse—cultural social life differs from the animal in kind. It is not just the expression of an animal of another kind. The reason why human social behavior is not organized by the individual maximization of genetic interest is that human beings are not socially defined by *organic qualities* [my italics] but in terms of symbolic attributes; and a symbol is precisely a meaningful value— such as "close kinship" or "shared blood"—which cannot be determined by the physical properties of that to which it refers. (p. 61)

My treatment of sociobiology in its possible psychological manifestations is all too brief, largely because this is an emergent rather than realized format of theory and research for psychologists. Because of some of its inviting features, I thought it appropriate to at least mention its potential relevance as a new type of organic orientation. It is not an orientation that this book shares, and I see in it problems common to all organic metaphors. In summary, let me elucidate some of the problems.

The use of the term *organic metaphor* in understanding human development has its source in the thinking of Aristotle (MacMurray, 1961). Aristotle's conception of the infant human was that it is an animal organism which becomes rational and acquires a human personality in the process of growing up (MacMurray, 1961). Our own cultural way of thinking has been deeply influenced by the Aristotelian way of thinking, and this use of biological analogies and organic categories has been strongly reinforced by the organic philosophies of the 19th century and the consequent development of evolutionary biology, of which sociobiology is its latest exemplar. Because of the large sweep of evolutionary conceptions and categories, they at first appear as appealing alternative outlooks to metaphors that are more mechanical in nature. Evolutionary conceptions are usually more synthetic than the traditionally analytic mode of thinking characteristic of mechanical metaphors. As a central ordering principle, the organic metaphor is peculiarly reticent on where ethics and the practical life are to be placed within its framework. In the end, it usually becomes an ethics of survival based on the assumption that all organisms adapt in order to survive. Although very sophisticated systems of ethics have been built on this premise, it nevertheless presents some serious problems for the guidance of our practical life. MacMurray (1961), in asserting the primacy of the personal as a central ordering metaphor for our understanding of the human world, has this to say of all conceptions that place superordinate emphasis on the organic conception of adaptation:

> We are not organisms, but persons. The nexus of relations which unites us in a human society is not organic but personal. Human behavior cannot be understood, but only caricatured, if it is represented as an adaptation to environment; and there is no such process as social evolution but, instead, a history which reveals a precarious development and possibilities both of progress and of retrogression . . . the personal necessarily includes an organic aspect . . . and this organic aspect is continuously qualified by its conclusion, so that it cannot even by properly abstracted, except through a prior understanding of the personal structure in which it is an essential, though subordinate component. A descent from the personal is possible, in theory, and indeed in practice; but there is no way to ascend from the organic to the personal. The organic conception of man excludes, by its very nature, all the characteristics in virtue of which we are human beings. To include them we must change our categories afresh from the beginning. (MacMurray, 1961, pp. 46–47)

When we speak here of the inadequacies of the mechanical and the organic as metaphors for our understanding of the human, we do not mean, therefore, that they are totally inappropriate and lacking in merit. No doubt, there are certain aspects of human behavior which can be understood within the prisms of mechanism or organism. As symbolic animals, human beings partly become the symbols they behold. For example, in serious psychopathological occurrences, people have been known to conceive of themselves as machines. I am saying that there is something about the human world which defies that characterization, and we are thus allowed to consider this type of psychopathological process as aberrant. I would contend with MacMurray (1961) that the use of mechanical and organic metaphors is, and should be, restricted in a very real sense. Where useful, they must be embedded in a more complex dialectical metaphor that treats the communicative process in humans as a unique and singular kind of human activity which should subordinate and never be subordinated to organic and mechanical language.

It must be understood that our choice of metaphors is first and foremost a moral rather than intellectual choice. In saying this, one does not preclude intelligence but rather includes cognitional activity as part of moral choice. Our moral choices assume that humans are not only made but also make themselves. In the absence of anything else, mechanical and organic metaphorical devices mask out important human ethical concerns. They are, as I have briefly demonstrated, lacking in critical intent and reflexivity. Until now, my own critical stance on various psychological metaphors leaves the reader with a vacuum. As such, critique does not fill the vacuum it creates. It is, therefore, incumbent upon critics to propose a positive program or alternative. At least, this is the position taken in this work. Accordingly, we move on to a thumbnail sketch of a metaphor we are calling the "personal."

THE PERSONAL METAPHOR

How can one call the personal a metaphor in distinction to the analogues of mechanism and organism? I will develop the idea that the personal metaphor is exemplified in the very process of human communication itself (i.e., language). Since language assumes reciprocity (i.e., communications are always between people), one can say that the unit of analysis is not an individual actor or monad but rather a relationship of *dialogue*. In other words, the "unit of analysis" is a totality (I–thou) that cannot be broken down further without losing the core metaphor of communication. For conventional psychology, this is a radical departure. This will be made clear as I go on. For now, let me offer a provisional definition for the position I am developing as that of a critical interpretation of the personal world that will demand a further explication of the personal world and an *emergent* role for the psychologist as critical interpreter.

The Personal World

The conception of personal world that I am developing must be clearly distinguished from a conception of the personal which is oriented toward the exploration of subjectivity. Much of phenomenological and humanistic perspectives are oriented toward the exploration of subjectivity. My contention is that this is a reaction, in the extreme, to objectivistic psychology (e.g., behaviorism). Without denying the importance of the subject and his or her consciousness, the position developed here is primarily and superordinately focused on language and its cultural functions. Emphasis on language reveals the importance of the relational quality of the personal world that I am attempting to develop. Some preliminary distinctions are therefore in order to prevent unnecessary confusion.

First of all, let me distance myself from what I will call the "politics of subjectivity." For psychology, this politic represents a countercultural vein in conventional psychology identified broadly as humanistic psychology. I situate it as a neoromantic revolt in the 1960s against the technical control rationality so prevalent in behaviorism. The politics of subjectivity issuing from this romantic impulse are well known in humanistic circles:

> Subjective feelings and states have become the only aspects of experience that can be trusted and that really matter. As the world around us appears increasingly remote and inaccessible, we seek refuge in self-absorption, contacted by an ideology that claims this to be the only route to self-fulfillment. This era of radical subjectivity is at once the logical extension of an unbridled individualism and a response to the hollowness of the social relations which a society based upon such individualism has brought about. (Cagan, 1978, p. 232)

The political impulse, generated by this neoromantic revolt against objectivity and its corresponding technologism, was a deliberate emphasis on private consciousness. This privatization of consciousness has been severely criticized as an inadequate alternative to technological rationality (see Cagan, 1978; Jacoby, 1975; Sennett, 1976). This preoccupation with the self lacks a historical dimension that I will attempt to develop as the cultural. Jacoby's (1975) critique of subjective politics is that it identifies the person as existing in a no man's land of free-floating interpersonal relations. I would agree that the critical path to the personal world lies elsewhere, penetrating the categories of individual and society, not merely juggling them. "The individual, before it can determine itself, is determined by the relations in which it is enmeshed. It is a fellow-being before it's a being" (Jacoby, 1975, p. 34).

That, in a nutshell, is the conception of the personal I will be attempting to develop. Our entry point will be language rather than subjectivity. With language preeminent, communication becomes an essential feature of this metaphor, and it reveals the fundamental *relational* quality of a metaphor based on "persons in relation" (MacMurray, 1957):

> Speech is public. It is at once thought and action, or rather a unity of which "mental" and "physical" activity are distinguishable but inseperable aspects; and as a result it establishes communication, and introduces the you as a correlative of the "I." For if

this "I think" logically excludes the second person, the "I say" is logically incomplete. To complete it we must formulate it as follows: I say to you, and I await your response. Thus the problem of the form of the personal emerges as the problem of the form of communication. (p. 74)

The peculiar task of a psychological discipline in this regard would be in the development of some form of systematic interpretation of these communicative expressions or human acts. We are, therefore, talking about an interpretative psychology (see Gauld & Shotter, 1977; Shotter, 1975). The whole issue of an interpretative psychology needs elucidation, and I will commence upon this essential task shortly.

The focus on language as opposed to subjectivity also calls for a shift in thinking as to what is to be the primary unit of analysis for our theoretical and research efforts. The conception of a personal world herein assumes that the primary unit of analysis is the person or the individual. At the same time the individual is not considered as an isolated monad. Our emphasis on the personal does not preclude larger structural entities (e.g., class structure, gender, etc.).

It serves notice that the personal expression is a fundamental entity within this framework and cannot be excluded for any research or theoretical experiences (e.g., Skinner and much of behaviorism). At the same time, this interpretation of individual expressions is not individualistic or egocentric in nature (see MacMurray, 1957; Sampson, 1977, 1981). We are assuming that the individual expressions or symbols are for purposes of communication; therefore, the unit of analysis must be a "you and I" rather than an isolated "I." Our assumption is that individuals use expressions because they intend meanings for others. Expressions are *ambiguous* because the meaning or significance of expressions is contingent upon multiple actors communicating with one another. Humans, it would appear, are peculiar animals involved in webs of significance that they themselves have spun (Geertz, 1973). "Webs of significance" is another way of saying that meaning is multiple or ambiguous, inviting interpretation. A human culture may be defined as a web of significance. Because of the intricate nature of human cultural expressions, it is accepted that the analysis of meaning within an interpretive perspective will not be adequately characterized with an experimental science in search of laws, but rather with a self-consciously interpretive search for meaning (Geertz, 1973). One of the persistent problems of interpretative approaches is their tendency to resist, or be permitted to resist, conceptual articulation and thus to escape systematic modes of assessment (Geertz, 1973). It is crucially important in the development of a self-reflexively interpretive psychology to resist that type of ambiguity which tends to obscure meaning. It is, therefore, necessary to attempt to spell out some of the parameters for an interpretive psychology so that it can be evaluated alongside other suggested approaches.

Interpretation. It has been suggested that psychologists consider themselves "hermeneuts" (Hudson, 1972), after the Greek god Hermes, the "interpreter." Interpretation is an activity that deals with the relational quality of communication. It is an attempt to understand the meaning of symbols. Interpretation attempts to make clear or give the sense of an object of study (Taylor, 1971). Normally it is considered a process that clarifies or renders the meaning of a text; but it can be extended to "text analogues" (e.g., a human person or community or culture) which, in some way at first sight, appears unclear, perplexing, confused, incomplete, cloudy, seemingly contradictory in one way or another. Interpretation aims to bring to light an underlying coherence or sense (Taylor, 1971).

Descartes. The *Discourse on Method* (1637/1960) is the most systematic attempt to eliminate interpretation. Paradoxically, it is an interpretative system nevertheless. In it, Descartes outlines four rules for achieving certainty and eliminating ambiguity:

> The first rule was to accept as true nothing that I did know to be evidently so; that is to say, to avoid carefully precipitancy and prejudice, and to apply my judgments to nothing but that which showed itself so clearly and distinctly to my mind that I should never have occasion to doubt it.
>
> The second was to divide each difficulty I should examine into as many parts as possible, and as would be required the better to solve it.
>
> The third was to conduct my thoughts in an orderly fashion, starting with what was simplest and easiest to know, and rising little by little to the knowledge of the most complex, even supposing an order where there is no natural precedence among the objects of knowledge.
>
> The last rule was to make so complete an enumeration of the links in an argument, and to pass them all so thoroughly under review, that I could be sure I missed nothing. (p. 47)

The first rule establishes methodological *agnosticism* or doubt as a principle for the acquisition of knowledge. Its contemporary counterpart would be the null hypothesis in statistics. The second rule establishes *atomism* as an ordering principle. The third rule establishes *hierarchy* as a principle of composition. And finally, the last rule establishes *induction* and *deduction* as superordinate logical procedures (for example, in distinction to some form of "dialectical logic").

Descartes' methodic doubt was to reflect a demand for certainty which accepted as true only those expressions or ideas that could be said to have achieved *clarity* and *distinction*. Ultimately, Descartes, the mathematician, found clarity and certainty in mathematical expressions. Clear and distinct ideas had no need of interpretation; their coherence or meaning would be self-evident. Descartes wanted assurance and a firm footing for knowledge. For the problem of interpretation is, basically, how we gain assurance that our interpretation of a situation is true or verifiable. By accepting interpretation as a real problem rather than denying it, as Descartes ultimately did, we are left with the fact that we understand expressions or ideas by referring to other expressions or ideas. In other words, we are left with what Descartes tried to eliminate: that is, the "hermeneutical or interpretive circle":

> The circle can be put in part–whole relations; we are trying to establish a reading for the whole text and for this we appeal to readings of its partial expressions and yet because we are dealing with meaning, with making sense, where expressions only make sense or not in relation to others the reading of partial expression depends on those of others, and ultimately of the whole. (Taylor, 1971, p. 6)

The circularity of this process has been a sore point where certainty has been the legacy of the Cartesian methodological synthesis. In this methodological world outlook, the method that led to truth and certainty was the one that was allied to propositions or events which were "clear and distinct."

As this method has developed over the centuries in all its guises, it has usually been allied with all methodological outlooks that eschew *ambiguity of expressions* and has attempted to relieve the embarrassment of conflict of interpretations by supplying a method which believes that it can achieve certainty (truth) over interpretation. In the context of our present discussion, this ideal of *certainty* demands that the "hermeneutical circle" must be broken out of because it is a vicious circle (Unger, 1976).

Empiricism and rationalism are two historical ways of attempting to break out of the hermeneutical or interpretative circle. Both these methods are descendants of the Cartesian mind–body dualism. Thus, Western thought may be said to be limited to a fund of basic schemes of explanations; ultimately they can, with all their variations, be boiled down to two ideal types—logical analysis (i.e., rationalism), and causal explanation (i.e., empiricism) (Unger, 1976). Both schemes provide an interpretation of what it means to account for something both in the sense of telling what a phenomenon or event is like, which is description, and in the sense of establishing why it had to follow from something else, which is *explanation* in the strict sense. Both these schemes appear historically in order to deal with the *interpretation* of physical and social events (Unger, 1976). The position being developed in this chapter proceeds under the assumption that psychology is, whether it likes to be or not, an interpretative science. Taylor contends that historically these two *schemes of interpretation* came to be what they were because of the problems presented by *conflicts in interpretation*. Rationalism is essentially the use of deduction. In deduction, certainty is achieved by grasping the inner certainty of propositions. The method of deduction is logical analysis, which is a linear development of arguments. Objectivity or certainty is gained by retreating to a "truth" that is formal and abstracted from concrete historical events. The conclusion in a deductive sequence is, therefore, a formal truth. The methodological thrust of rationalism is to bring understanding (i.e., clear and true interpretation) to an inner clarity that is absolute (i.e., logical argument). Rationalism provides an ordering principle that exists at the level of ideas (Taylor, 1971). It is essentially an ahistorical method of arriving at certainty. Empiricism is a method for arriving at truth at the level of events in history. In contrast to strict rationalism, its truths are time-bound and concrete. It argues linearly at the level of cause and effect, the cause in a time sequence being antecedent to the effect.

Empiricism attempts to break out of the circle of interpretation by attempting to move beyond the problem presented by human subjectivity. Empiricism as a method systematically attempts to reconstruct knowledge in such a way that there is no need to appeal to readings or judgments. For this reason the basic building block of empiricism is the sensory impression or sense datum that psychologists have come to know as the stimulus. At the level of perception, the stimulus is a unit of information that is supposedly not mediated by judgment (Peirce, 1934). Subjectivity is played down within this formulation by alluding to the organism as a *tabula rasa* (Ausubel & Sullivan, 1970; Ausubel, Sullivan, & Ives, 1980). At the sensory level of *brute datum*, it is therefore perceived to be unnecessary to utilize subjective interpretation or judgment. Empiricism as a methodological ideal wants certainty anchored beyond subjective intuition.

One other important feature in the quest for certainty is that its ultimate test is to be found in the individual consciousness. In Descartes' time, this was tantamount to a rejection of scholasticism, which saw truth as resting on the testimony of sages and of the Catholic Church (Gadamer, 1975a). In its more generic form, this was the rejection of historical tradition for an ahistorical alignment of truth within a contemporaneous individualist consciousness. In retrospect, one can see the Cartesian legacy as the beginning of the methods of modern natural science. This was a historical circumstance, which is too often forgotten. The legacy of Descartes represents a bridge into modern consciousness that is exemplified in the qualification of the scientific world view. Gadamer articulates the historical movement of a scientific empiricism. The importance of scientific method consists in the fact that, by its own definition, it investigates only by elaborating questions accessible for mathematical abstraction and measurement or its equivalents (Hodges, 1969). One could say, then, that this is the Cartesian legacy.

Vico. In spite of the momentum of a world view wedded to methods from the natural sciences, Descartes had a contemporary, Vico, who was to propose an alternative, based on historical methods, to the natural science methods in the study of the human. In his treatise *The New Science* (1725/1970), Vico attempted to outline a "science of the human" based on *history* rather than *nature*. The "new science" was to be peculiar to human nature: a science of humanity whose dominant framework was to contrast in some crucial ways with the Cartesian synthesis (Belaval, 1969). He concluded that an approach to human studies could not be made by methods of detached observation but only by *understanding* and *interpretation*. Thus, events in the human sciences would be understood *sympathetically* rather than from a position of *detached observa-*

tion (Berlin, 1976). Vico criticized the Cartesian synthesis because it, in its methodological outlook, systematically obliterates the importance of personal and cultural history which is, in essence, cultural memory. The role of the image (i.e., imagination) is also curtailed by the Cartesian critical spirit (Gadamer, 1975b). It was Vico's conviction that Descartes' criteria of "clear and distinct" ideas could not be properly applied outside the fields of mathematics and natural science (Rickman, 1969). Systematic use of historical methods would clearly be needed to explore the full range of human capabilities and the institutional life that ensued from these capabilities. Sympathy as identification was contrasted with the Cartesian method of doubt. We are able to know and identify with persons and institutions because we are part of their makeup. To interpret and come to understand human institutions, it would, therefore, demand a more anthropological perspective at an institutional level and a use of biographical and ethnographic methods at the level of the personal.

The legacy of Vico from a historical perspective follows a different historical line from the Cartesian one. Gadamer has traced these polar tensions into the latter part of the 20th century from a historical perspective while at the same time providing a strong rationale for use of interpretive (hermeneutical) methods. He essentially resurrects the issue of a "human science" in his attempt to understand historical texts. He draws his impetus from the 19th-century German version of the Descartes–Vico tension, which was directed toward the issue of how there could be a truly human science. These issues are discussed in much greater detail and acumen in his masterwork, *Truth and Method* (Gadamer, 1975b). Suffice to say, he picks up on the tradition in German thought, pioneered by German 18th-century historians, which advocated a science of humanity based on sociohistorical reality made meaningful because it consists of expression of human thoughts, intentions, feelings, and valuations. In the tradition of Vico, this line of thinking provided a rationale for studying phenomena as autobiography and biography, where methods used by historians are most appropriate (Gadamer, 1975b). In short, we see here the contrast of the classical world view with that of the hermeneutical or interpretive.

Although these polar tensions have been experienced in continential European thought, especially German thought at the turn of the century, the development of the discipline of American psychology appeared to have sidestepped this antinomy. American psychology brushed over this controversy because of its easy embrace of the scientific and technological perspective.

Normative Science. The antinomy between Descartes and Vico represents real historical moments that are conflictual in nature, concerning conceptions of truth, objectivity, and rationality and the superordinate method of arriving at what is true (is) and what is good (ought). The Cartesian synthesis operates out of an epistemological horizon: How do I know that I know? Vico represents the opposing horizon of starting from the horizon of normative categories: How do I know what to do? Here the philosophical horizon is viewed from the stand point of the practical. In using the term *practical* here, we do not mean to convey the sense of a technological pragmatism. How I know what to do is not simply a pragmatic consideration. Using the question of "what to do" signifies a priority given to the normative (ought) over, but not excluding, the epistemological concern (regarding what is).

The shift to a psychology ventured from a normative standpoint is no mere conceptual shift. It involves substantial changes in practice which, at present, are unfamiliar territory for psychologists trained in the "mainstream." Shotter (1975) dramatizes this point:

> The new psychology of action we propose requires, thus, not just a change of content compared with old, but a radical new form of thought and mode in investigatory activity. It aims at producing not publicly shared objective knowledge, but intersubjectively shared understandings, not discovered by individuals searching in the world above, but made and agreed upon by people working in dialogue together. (p. 135)

It is clear that the dominant interest of psychology as a discipline, up to this point in its short history, has been weighted toward the epistemological horizon. The position I will ultimately articulate in this book will reflect a shift to the normative; that is, the overriding initial concerns will be seen from the standpoint of action. In order to clear the ground for this, it will be necessary here to figure that we are shifting horizon or standpoint (Lonergan, 1957). The type of inquiry I am suggesting is, in a sense, a different kind of science. Habermas (1972) characterizes part of the shift I am taking as historical-hermeneutic:

> The historical hermeneutic sciences view knowledge in a different methodological framework. Here the meaning of the validity of propositions is not constituted in the frame of reference of technical control. The levels of formalized language and objectified experience have not yet been divorced. For theories are not constructed deductively and experience is not organized with regard to the success of operations. Access to the facts is provided by the understanding of meaning, not observation. The verification of lawlike hypotheses in the empirical–analytic sciences has its counterpart here in the interpretation of texts. Thus the rules of hermeneutics determine the possible meaning of the validity of statements of the cultural sciences. (Habermas, 1972, p. 309)

Cultural Science and the Personal World. Habermas (1972) above uses the appellation "cultural sciences." This is most appropriate in the conception of the personal world as will be developed in this work. From the perspective that I will be developing, the personal world is the cultural world; culture is the womb in which personhood grows and transforms. When I say grows and transforms, I am not speaking from the point of view of an organic metaphor. When one speaks of personal growth, that is saying that the personal world is cultivated within cultural forms. By cultural forms, I mean specified lived histories which constitute the relational totality of the personal world. Thus, the relational totality of the "you and I" confirms that the personal world is cultural from the ground up. For me the notion of the personal world as cultural means that personhood is not an abstract entity. My personality expresses my cultural history. To know me, it will be necessary to know a certain species called "male Irish-American Catholic." That appellation alone covers a multitude of evils and many strengths. I'll speak no more of either, but the reader can take the liberty of reading between the lines as my position develops.

Critical Interpretation. The cultural world of the personal is the world of the stuff which makes for biography and ethnography. It is the world of the lived experience that phenomenologists speak of in their writings. To stop at this point is a radical departure from mainstream psychology. Much of what constitutes phenomenological psychology remains here. We must go further. Culture always exists in dynamic social relations of power. To ignore power relations means that such structural concepts as class structure, gender structures, and so on can be sidestepped in our study of the personal world. A psychological inquiry will be critical if its interpretative system encompasses relationships of power. To deal with power relations, one must introduce structure concepts (e.g., class structure) which are dialectically related to the cultural dimensions specific to distinct but related cultural worlds (e.g., Irish and English). Specifically, the reader will be engaged in a dialectical synergy of the principles of *analysis* and *synthesis*. The analysis of culture, which is the pole of the personal world, is set off and polarized with the structural dimensions of class structure, gender structures, race, and so on. In other words, it is understood at the outset that the personal phenomenological world of actors within cultures is embedded in larger structural totalities of class, gender, race, and so forth which involve relationships of power. The role of critical interpretation is to draw attention to these power relations and bring to light inequities of power. The interest of critical interpretation is *emancipatory*; that is, it challenges the stases of social relations where there are gross inequities in power. The task of critical interpretation, although epistemological, is first and foremost normative. When Marx (1844/1963) said that the role of philosophy is not simply to interpret the world but to change it ("11th Thesis on Feuerbach"), he was, in essence, saying that philosophy should be a normative enterprise. The same can be said for psychology as a social science endeavor. For most of its short history, it has hidden behind the concept of "value-neutral" inquiry in sidestepping the normative dimension. Value neutrality can no longer be sustained as a value position since it has become increasingly clear that it is an ideology sustaining a technical–control rationality (Habermas, 1970). Here, *ideology* means a suppressed interest that is either denied or not seen. Technical–control rationality activity sunders the distinction between the practical and the technical and represses the ethical dimension within it (Habermas, 1970). Depoliticized at the level of theory and theory language, the effectively neutral position masks political intentions. A critical look at behaviorism (Danziger, 1978), structuralism (Buck-Morse, 1975; Gardner, 1972; Sullivan, 1977a, 1977b, 1980a, 1981; Venn & Walkerdine, 1978; Wilden, 1975), psychometric psychology (Hunt & Sullivan, 1974; Kamin, 1974) and sociobiology (Caplan, 1978) gives ample evidence that social science theories are intricately involved in value issues and have broad social policy implications. At the level of superstructure, social science theories are "legitimators" of the status quo, that is, they render interpretations which back up or legitimate a certain sociopolitical constellation of power (Habermas, 1970). The position developed here makes no claims for being value-neutral. We hope that it systematically serves an emancipatory interest by the development of what we are calling "a critical interpretative science of the personal." The question now arises, how does criticism relate to interpretation? Bernstein (1971) locates the line of thinking for a critically interpretative social science in Marx and Hegel:

> It has the power not of delineating some utopian ideal which is to be striven for, but revealing to men a critical understanding of what they were suffering. Unless criticism does this, it becomes idle speculation; the test of the correctness of a radical critique is its ability to bring genuine human problems suffered by man to a self-conscious human form. (Bernstein, 1971, p. 54)

Although my own treatment of critical interpretation does not eschew the utopian, I nevertheless agree that one important aspect of critical interpretation is to bring to awareness (i.e., interpretation) systems of power relations under which people suffer. This makes critical interpretation a *normative* endeavor. The primacy of the *normative* over the *epistemological* is a radical shift in thinking from Descartes and Kant (Bernstein, 1971):

> It is not by any superficial or careless reasoning that most modern thinkers have been led to maintain dichotomies between the is and the ought, the descriptive and the prescriptive, fact and value. In a variety of ways, philosophers have argued that whatever status we assign values, norms, and ideals, they are not objective phenomena to be discovered in nature. Science, our most powerful and successful means for exploring nature, can tell us only what is, it can describe, explain, predict, but cannot tell us in any categorical sense what ought to be. If we want to find the philosophical arguments in support of such a position, we need only study the works of most of the major figures or movements in philosophy since Descartes. Hume (in some of his moods), Kant, classical materialism, logical positivism and empiricism are all agreed about this "dogma." But it is precisely this dogma that is the focal point of Marx (and Hegel's) attack. (p. 71)

From the Greeks to Descartes, we have clearly witnessed a historical shift moving from the Greek preoccupation with the ethical to a preoccupation with epistemological categories. By the time we reach the 18th century, these categories or dimensions of human existence (i.e., the true and the good) have developed an almost separate status from one another. By the time that Kant comes to deal with these dimensions of human life, he is able to treat them under three separate critiques. His *Critique of Pure Reason* essentially deals with epistemological categories and their truth status. His *Critique of Practical Reason* treats separately the category of the ethical. His other critique, which we will not discuss here, deals with the "esthetic." The important point we wish to note about this great thinker is that by the time we have reached his synthesis, one sees the almost complete separation of ethics from epistemology. This is not to say that, in the case of Kant, the ethical is bypassed. For this great thinker, the practical was close to his heart, so much so that he would write a whole critique on the importance of ethical questions. By the 20th century, the nascent field of psychology can abandon the ethical category and claim, as with other social sciences, that its task is value-neutral.

Modern psychology, for the most part, completely sidesteps the ethical dimension in its quest for epistemological certainty. This quest unites very strange bedfellows indeed. Hull, Piaget, and Chomsky can all fall under the same umbrella if one considers that all of their theories are preoccupied with the primacy of the epistemological dimension. Even though theories in contemporary psychology differ in some essential ways, they are brought together under one roof by their relative exclusion of the normative dimension. With the primacy of the epistemological category, we also see the relative emphasis of theory over practice. By the time we reach the 19th century, all these are dichotomized in thought and institutionalized in the buregoning social science disciplines. For the social sciences, the development of good theory becomes a raison d'être. They will proudly stand aside and leave the practical life to the philosopher and the lay person.

It may be said without hesitation that psychology, as a disciplined social science in the 20th century, is pre-Hegelian and pre-Marxian in its philosophical predispositions (Sève, 1978). This does not deny that psychology has conflictual elements within this pre-Hegelian synthesis. One of these conflictual elements can be seen in the distinction between mechanical and organic metaphors already discussed in this chapter. In fact, psychology in the 20th century has been caught philosophically between Hume and Kant. To put it another way, it has been caught between mechanistic empiricism and some form of constructive organicism. The noise created by this conflict has masked its pre-Hegelian origins. My proposed orientation to the practical, in effect, shifts the ground to a different level of argumentation and thus to a different level of practice in the discipline. The metaphor of the personal, which we are developing, self-consciously appropriates the dimension of history as an essential ingredient in psychological inquiry. Philosophically, the metaphor of the personal is therefore indebted to both Hegel and Marx. The centrality of history cannot be underestimated when making this change in perspective. In comparing Kant to Marx, Bernstein (1971) emphasizes some crucial differences that make their perspectives tantamount to different points of departure:

> Marx would agree that all observation is "theory laden" and that the reality we know
> and encounter is conditioned by the "forms of life" that have evolved in human social
> institutions. . . . . The central motif here can be traced back to Kant who emphasized
> the constructive, categorical aspect of judgment. But unlike Kant, Marx does not
> believe there is any Ding-an-sich that stands apart from human knowing, and further-
> more, Marx would maintain that our basic categories change and develop in history.
> Nor is Marx sympathetic with those thinkers who interpret man's cognitive perspective
> as an act of individual will or arbitrary convention. What is distinctive about Marx's
> reflections on human cognition is the way in which he relates it to the evolution of
> man's practical needs as manifested in his social life. It is praxis that turns out to be the
> key for understanding the full range of man's developing cognitive activities. (Bern-
> stein, 1971, p. 73)

This shift in horizon initiated by Hegel and Marx is developed in contemporary form by Habermas, working within the broad tradition called the Frank-
furt school. In *Knowledge and Human Interests* (Habermas, 1972), the traditional conception of rationality as "value neutrality" is challenged and put in
historical context.

> In this field of inquiry, which is so close to practice, the concept of value-freedom (or
> ethical neutrality) has simply reaffirmed the ethos that modern science owes to the
> beginnings of theoretical thought in Greek Philosophy. Psychologically an uncondi-
> tional commitment to theory and epistemologically the severance of knowledge from
> interest. This is represented in the logic by the distinction between descriptive and
> prescriptive statements, which makes grammatically obligatory the filtering out of
> merely emotive from cognitive content. (Habermas, 1972, p. 303)

Habermas (1972, 1974) introduces the conception of "knowledge constitutive interests," which are by definition shapes of what counts as the objects and
types of knowledge. As a Marxist, Habermas claims that these interests are basic
because they are rooted in specific fundamental conditions of possible reproduction and self-constitution of the human species (Habermas, 1972). Although the
epistemological status of these interests is open to question (see Bernstein, 1978;
McCarthy, 1978), I nevertheless introduce them here because they are suggestive for the development of my present work. Three primary cognitive interests are articulated by Habermas (1972): the technical, the practical, and the
emancipatory. Three types of sciences are articulated corresponding to these
interests:

1. Empirical–analytic sciences
2. Historical hermeneutic sciences
3. Critical emancipatory social sciences

The empirical–analytic sciences, with their emphasis on technical control,
are evident in most examples of mechanical metaphors that we have given. The
conception of the personal I am attempting to develop incorporates a her-

meneutical dimension but at the same time attempts to be emancipatory and critical in nature. It is hermeneutical in nature because it systematically takes as its task the problems posed by the interpretation of symbols. The peculiar task of a psychological discipline in this regard would be in the development of some forms of interpretation to deal with individual life histories embedded in cultural forms. The position that I have been developing as an interpretive psychology starts out with the assumption that individual selves are agents whose actions intend meaning. In contrast to an American ideal that has treated the self as a self-contained individualistic system (Sampson, 1977), I am proposing a unit of analysis that assumes persons in relation (MacMurray, 1957). The unit is a collective, the you-and-I.[5] Meaning is a constitutive element in a relation of I–thou. Therefore, meaning is not simply that of individual actors but is part of the *expressive* relationship between actors.

This is an important shift in perspective and should be noted, since the ontology of most of mainstream social science in the psychological domain lacks this minimal communal perspective (Taylor, 1971). The considerations we are discussing here are currently being pursued by a network of researchers who conceive of their work as a phenomenological approach to psychology (Giorgi, 1970; Giorgi, Fisher, & Murray, 1975). There is much in this phenomenological line of attack with which we are in broad sympathy. Giorgi's (1970) work, for example, is most helpful in shifting the ground away from mechanism. There is, however, a significant omission in this phenomenological emphasis; this is discussed under Habermas's (1972) third cognitive interest—the critical emancipatory interest. This third interest is fundamental to our development of the "personal" and, I believe, allows me to distinguish my position from straightforward phenomenological viewpoints. This accounts for our use of the term *critical* as one of the terms in the title of this book.

A psychology that is "critically interpretative" rests on several assumptions about the whole process of interpretation. First, a critically interpretative psychology is reflexive or self-reflective in nature. It is understood that psychological interpretation is embedded in history and, therefore, that the psychological interpreter must appropriate the responsibility of operating from a biased viewpoint. She or he is prejudiced. My understanding of bias and prejudice does not assume that this situation should be eliminated. A self-reflexive psychology simply takes the responsibility for having a viewpoint or a set of preunderstandings (regarding prejudice). Self-reflexivity in interpretation is simply accepting the reality of viewpoint, bias, and so on, and dealing with the problematic nature of same.

---

[5]The unit you-and-I is the most primitive term and cannot be reduced. In saying this is a minimum we are saying that the you-and-I can encompass more than simply two people but never less. As a primitive term, the you-and-I indicates that persons are persons in relation. For me, *persons in relation* is another way of saying *culture*. This will be developed in greater detail in Chapter 3.

One assumes that psychological interpretation is embedded in a societal history and not beyond it. Whatever universality that interpretation achieves, it is a universality that is bound by time and space and, therefore, exists to be transcended. Thus, the conception of truth within this context is dynamic rather than static. Second, a psychology based on a critical interpretation of the personal world is fundamentally interested in the concept of freedom and free acts of individuals and groups. In fact, our very definition of the personal world assumes a relational freedom. Thus, instead of considering interpretation to be disinterested, our own conception embodies an emancipatory interest (cf. Habermas, 1972). This conception of psychology is a part of an emancipatory social science. Its uniqueness in contradistinction to conventional social science is its expressed cognitive interest in *emancipation*. The self-reflection of a critical inquiry brings to light social relationships within society based on power and sustained by inequities of power (regarding domination). The emancipatory cognitive interest brings to light these inequities of power (e.g., between men and women) and attempts to provide a system of self-reflection that will end in a transformation of these relationships of domination. (Fowler, 1974)

Psychology, like most social science disciplines, is derivative of social and economic structures in a society. For the most part, what is considered mainstream psychology is a symbolic legitimation of the economic structures in the society (i.e., the social relations embedded in late capitalism). In a nonreflexive way, it has been the apologist for relations in late capitalism (i.e., ideological). In many instances it helps to defuse emancipatory impulses by "blaming the victims" of a particular economic order (Ryan, 1971). Relating an emancipatory interest to a critical conception of the personal, one sees the ascendance of a set of relations between persons and groups that are based on a communication which is not premised on domination and oppression. A practical emancipatory interest "extends to the maintenance of intersubjectivity of mutual understanding as well as to the creation of communication without domination" (Habermas, 1972, p. 113). Of this we will say more. Briefly, a critically interpretative psychology announces social structures which build up a personal world and human projects and denounces, through its interpretations, social structures that destroy the personal as we conceive it.

PROSPECTUS

The discussion of psychological theories as metaphors was intended to locate, in a critical manner, the point of view of a critical interpretative psychology that I am about to develop. Although critical of other psychological positions as metaphors (e.g., mechanical and organic metaphors), I am nevertheless completely aware that my position is not a replacement for these endeavors. No amount of criticism will make a theory or practice go away when it serves certain societal interests. In addition, these positions are embedded in historical intellectual traditions which give them solidity and longevity. The precariousness of my position is that it has no historical roots in the broad discipline of psychology as it is presently understood. In fact, psychology as a discipline is unique from all other social science disciplines in that it completely lacks a critical perspective. Sociology, anthropology, political science, theology, and so on have substantial scholarly interests that support what I would label "critically interpretative" inquiry. Because of this vacuum in the psychological disciplines, I risk the possibility of being written out of most of the discipline's consensus. Many will say I am doing anthropology, sociology, political science, and so forth. I would contend that psychology needs the presence of a critical emancipatory interest and that it must, of necessity, be interdisciplinary in nature. I can only hope that there will be a significant number of psychologists who see the necessity of the point of view I am trying to develop under critical interpretation. If this happens, I will be satisfied.

The following chapters build on one another dialectically. Therefore, consecutive reading of the chapters is in order. The sketch of the personal metaphor in this chapter was thumbnail. By the end of this work, I hope to have developed an in-depth discussion of the personal world and a need for a critically reflexive psychological interpretation to understand it and to help transform it.

2

HUMAN EXPRESSION
A Relational Act

PROLOGUE

A psychology that is self-consciously interpretive and critical will follow the form of what I am calling the personal metaphor. One of the fundamental assumptions of this point of view is that human behavior is understandable only in terms of a dynamic social reference, the purely isolated self being a social fiction (MacMurray, 1957). The peculiar task of a psychological discipline, in this regard, is in the development of some form of systematic interpretation of communicative expressions, which I hold as synonymous with the term *human act*. I am assuming that individual or collective expressions are symbols for purposes of communication. Therefore, the fundamental unit of analysis must be a dyad (you and I) rather than an isolated individual or monad. The contention here is that individuals use expressions because they intend meaning for others.

EXPRESSION

The dictionary defines the term *expression* as an action of pressing out, to press out an act, process, or instance of representing or conveying in words or some other medium. It also notes that expressions are "signs or tokens." Here the term *human expression* in the chapter title indicates the sign quality of human action. This, of course, is in keeping with our emphasis that human expression is essentially and superordinately linguistic in nature. *Relational act* signifies that human expressions are espressions for others. They are communicative acts. The case can be made that all animals are expressive, and we would not argue against that notion here. From an evolutionary perspective, one can say that, in the broadest sense, animal expression is a particular coordinated set of movements that in some way manifests an interior being externally (Merleau-Ponty, 1963).

Tropisms, or certain stylized actions of animals, can be said to be animal expressions peculiar to different species of animals. One thing that will be argued in this chapter is that human expression cannot be subordinated to general biological expressions or mechanical acts. The emphasis in this chapter is to focus on the peculiar characteristics of the human act from the point of view of the actor or agent. This, of course, is a partial perspective. Since human actions are being treated as communicative expressions, our treatment is, of necessity at the outset, limited and incomplete. The following chapter will complete our conception of the human act as relational. We start our treatment by making some crucial distinctions that have already had important consequences in the short history of psychology as a discipline.

The Intelligibility of Human Expressions

Let us return to the distinction we made in our introductory chapter; that is, the distinction between the mechanical, organic, and personal orders. We will be calling these orders *modes of intelligibility* in this chapter.[1] We use the term *intelligible* to mean that human action forms meaningful complexes that are interpretable. An action or set of actions is interpretable if it can be said to fall under some meaningful schema of interpretation. An action is said to lack intelligibility if it cannot be said to fall under some schema or mode of interpretation. An action in an instance of this kind is judged meaningless. It is important here to emphasize the relational quality of intelligibility. An action, in the end, is not intelligible only from the point of view of the agent but is also dependent upon the interpretive schema of those who receive the expression to interpret.

Mechanical Mode. The mechanical mode of intelligibility rests on the assumption of the causality of physical systems. This assumption is that the *cause* of something *precedes the effect* in a process understood to be in temporal contiguity. What I have called "mainstream experimental psychology" rests on the possibility of explaining human behavior within the parameters of a causal process. The mechanical mode of intelligibility, therefore, assumes that human action or behavior can be made intelligible by the attribution of causality between a behavior and a set of antecedent conditions. The independent variable is the antecedent stimulating condition for the dependent variable or response. Usually, the attribution of causality was carried out under highly controlled conditions that specified in advance the antecedent conditions to be manipulated and the *response* to be evaluated or assessed. Within this mode, the response (R)

---

[1]The distinction we are using is made by John MacMurray in *The Self as Agent* and *Persons in Relation*. This distinction is not exclusive to him. Using different terms (i.e., physical, vital, and human orders), Merleau-Ponty presents a similar trilogy in his *Structure of Behavior* and *Phenomenology of Perception*.

was the behavior being assessed as a result of the experimentally controlled manipulation of antecedent stimulus conditions (S). Thus the appellation "S–R psychology." All this rests on several assumptions about the nature of the human act. The first is that human action is a mechanism like other mechanical processes in nature. To see human action as a mechanism in nature is, in essence, to say that human action has no unique status in the natural world. The second assumption rests on the principle of analysis already discussed in the first chapter, and that is that human action can be understood in terms of its more simple components. Following from the preceding assumption is a third, ancilliary assumption, that of reductionism. One assumes here that human behavior can be reduced to biology, biology can be reduced to physiology, and physiology can be reduced to inorganic physical systems. This accounts for the easy embrace of machines as an analogy for making human behavior intelligible. It also explains why animals could easily be substituted for humans for experimental convenience. The fourth major assumption is that all organisms are *patients* as opposed to *agents* in nature. It must be understood that this assumption is one of degree, with considerable variation between psychologists who endorse causal determinism in human behavior.[2] Nevertheless, it may be said that the organism or subject within the experimental setup is a "patient" in the whole process; that is, he or she or it undergoes the program of the experimental procedure. Here we are making the distinction between a patient who undergoes or receives a treatment and the agent who initiates or transforms his or her (or its) conditions. This distinction is important as we consider the ultimate *possibility and feasibility* of understanding human behavior as causally determined. Within the purview of mechanical intelligibility, the free act based on the intention of an agent is considered as irrelevant or as a fiction.

Historically, considering the prominence of experimental psychology, these assumptions have profoundly influenced the brief history of psychology as a discipline. This can easily be seen in the vast majority of what are called "learning theories" (cf. Hilgard & Bower, 1966). Although most early learning theorists of the behaviorist variety recognized different species capacities, their main emphasis remained on the mechanical and repetitive nature of most human as well as animal acts. This can easily be seen by looking at the way these theorists

---

[2]This is clearly seen in the extremes in behavioral psychology as exemplified in the work of Skinner as opposed to the work of, say, Hull and Tolman. The latter, with their intervening variables (e.g., habit, drive, etc.), attempted to deal systematically with some dispositional properties of the organism in the production of behavior. Skinner eschews the necessity of this way of theorizing in his essay "Are Theories of Learning Necessary?" Focusing on Hull for the moment, one can easily see that his intervening variables do not approach the concept of agency that we will be developing. Hull and Tolman always see these intervening variables as linked to the manipulated conditions of the stimulus. The intervening variable is behavioral psychology's closest concept to what might be called "a history."

treated such concepts as insight and understanding. For the moment we are treating these concepts from the point of view of ordinary language. In Thorndike's early work, we see the role of insight present but minimized. In Thorndike's system, insight simply grows out of earlier habits. In contrast, Pavlov finds no use for the term in his system, but one would have to say that his "second-signal system" is an attempt to come to terms with some of the factors related to insightful behavior. Guthrie is only derisive toward the use of any concept that approaches conscious understanding and insight; in Skinner, the term does not occur (Hilgard & Bower, 1966). Hull, Spence, and Osgood attempt to deal with these concepts in terms of "mediational processes" or intervening variables. The process is always linked to previously formed habits. Tolman is the only major behavioral learning theorist to employ the term *insight*, and he can be considered an isolated case, an exception to the rule (Hilgard & Bower, 1966).

The inability to see the organism and the human person proper as an active organizing power limits the scope of our understanding of human behavior. I would not be so foolish as to venture that we are developing a full conception of human nature and its action possibilities, but I nevertheless contend that the mechanical mode of intelligibility is too restrictive in its scope of what constitutes human powers, to the point of denigration. I am not here disagreeing with the fact that much of human behavior is habitual and therefore determined. I am saying, however, that mechanistic accounts of human behavior seriously underestimate some important and unique characteristics of human functioning by their failure to countenance such concepts as conscious human intentions, insight, intuition, and also unconscious intentions. These are only examples. The implicit fear here is that if such concepts as these were allowed their appropriate hearing, the ultimate possibility of explaining human behavior in terms of causally determined processes would be seriously challenged. The proponents of mechanistic intelligibility rightfully assume that by seriously entertaining these processes they would be abandoning a natural science of human behavior. I would venture that a natural scientific account of human behavior must be abandoned for a more complete understanding of the scope and variety of human functioning—a view that is readily obvious at a commonsense level of observation. If traditional learning theory has laughed at commonsense notions of human understanding as active conscious agents, the "man on the street" has returned the compliment by treating accounts of traditional learning theory as unbelievable.

The Organic Mode. To talk about organic intelligibility is to attempt to understand human behavior at the level of biological systems versus simple physical systems in nature, such as the mechanism. Hence a shift in emphasis from physics to biology. This does not necessarily detract from an interest in habitual behavior patterns as seen in the mechanical mode. To the contrary, biological interpretations see the production of habitual behavior patterns as species-specific adaptation. One, therefore, looks for different habit patterns between species. For this reason, in contrast to the mechanical mode, organic intelligibility is not reductive in nature.

Second, from the point of view of organic biological systems, the organism is seen in an *active* interchange with its environment, in contrast to the reactive position of mechanism. The conception of an active organism as a living system interacting with its environment brings the notion of teleology into consideration. *Teleology* is a term that captures the possibility that organic actions are goal-oriented and directed toward end states. Explanation in terms of a teleological framework substantially shifts the ground from the causal mechanisms exemplified in the mechanical mode to explanatory frameworks that have goal-directedness as part of the explanatory system. Taylor (1964) sees this as a crucial difference between teleological explanations and mechanical explanations, because the former require the use of a premise that violates the formal requirements for mechanical explanations. Teleological explanations assume the fundamental adaptive importance of ecological relationships in end states.

The most important characteristic of organic adaptability is that the teleology involved in adaptation is a habitual pattern that does not constitute consciousness and human intention. This distinction is important when we later consider conscious intentions. The various manifestations of structural conceptions in cognition (Piaget), linguistics (Chomsky), and morality (Kohlberg) are paradigm cases of organic modes of interpretation. The important point to be made about all these cases is that the understanding of a phenomenon is not primarily in terms of the demonstration of a causal sequence, but in the articulation of the *form* or *structure* that generates phenomena (content). This is called "the form/content distinction." All these theorists have criticized mechanistic explanatory systems as essentially involved with surface phenomena. The uniqueness of these systems of interpretation is that they seek "intelligible forms" through a process of abstraction from the phenomena that they are studying (i.e., language samples, moral judgment protocols) and then attribute explanatory power to the forms that are articulated. For example, Chomsky talks of a "generative grammar" that is capable of producing an infinite sequence of linguistic phenomena.

Our intention here is not to issue an elaborate critique of these systems of interpretation but to establish clearly how they differ in some essential ways from the position I am developing. For all their merit (and I say this with a sense of admiration), these systems of interpretation or intelligibility systematically sublate the importance of the role of conscious intentions of the human agents. By this I mean that these systems avoid this dimension as essential in their explanatory frameworks. All these systems ultimately subordinate the importance of *self-consciousness* into larger biological systems.

Finally, emphasis on *active structures* attributed to the organism proper as a historical, universal, abstract form seriously underplays the role of the relational quality of human action. The active forms or structures attributed to the organism are ultimately individualistic in nature. It is not that these theories lack a relational historical element. Rather, the element is so totally subordinated to the emphasis on finding structures inherent to the organism that the historical resistance factor (i.e., the environmental order in which the organism operates) is a stage prop that is the background for action. Our attribution of individualism is a judgment of degree. We say these theories are individualistic because they clearly articulate the structure of the individual while the structure of the environment is left unspecified in the background. At the level of theory, all the structural theories are relational and interactive. At the level of practical consequence, these theories fall far short of the mark of clearly and systematically coordinating organism and environment as correlatives in an adaptive biological process. [3]

We are now about to enter into a discussion of the personal mode of intelligibility. A word of caution is in order. We have not discussed the previous modes of intelligibility simply to dismiss them and, like Dante's Beatrice, bring the reader into Paradise. The reader would simply be misled about my intent. To move to a human order of intelligibility is not to abandon the previous orders. Humans are more than physical mechanisms or biological entities, but at the same time they are that all the same. For example, the nerve impulse is clearly understood at one level of interpretation as a biochemical mechanism (i.e., synaptic transmission). At the same time, we are saying that you cannot move from this level of interpretation and fully comprehend factors that we call characteristically human. In fact, these biological and physiochemical orders may be better known when they are understood and subordinated to a "human level of interpretation" (Merleau-Ponty, 1963). One hopes that the position being developed will make this clear. For now the reader will have to take a promissory note that clarity will ensue.

The Personal Mode. The unique quality of the personal mode of interpretation is that the system of interpretation systematically tries to understand the person as an active agent whose expressions are for others. The unit of analysis, as MacMurray (1957) sees it, is not the I but the you-and-I. This is not a trite distinction. Most personality theories as I understand them are egocentric; that is, they exclusively focus on the I and collapse the dialectic of the you-and-I. When I speak of the personal, I am not just speaking of dyadic relationships. The

---

[3]Our own treatment will eventually differ from the analytic tradition's discussion of reasons as teleological explanation. We will follow MacMurray's (1957) distinction between teleological mechanism and the concept of intention. We will try to demonstrate the importance of this distinction as an essential differentiating characteristic between organic and human orders of intelligibility.

unit of the you-and-I expresses the fact that all human expressions are *correlative* expressions between persons. Therefore all human acts are *communicative acts.* All this needs now to be elaborated systematically in this section and the subsequent chapter. In this chapter, we will start out by discussing the *human act.* In the next chapter, we will start out by discussing the human action by discussing it under the appellation *personal world.*

Let us start out by examining how we are going to use the expression "the human act." I am assuming that "the human act" is a significant expression that is directed beyond itself to other selves. Biological acts such as defecation are not considered human acts because they are shared by humans with other species and are not typically directed toward other selves as signals. I call a human act an act—rather than a movement as a behavior—because of the intentional quality of human action. This will be expalined shortly. Further, the human act is bipolar or dialectical in nature; that is, the forward movement of dynamism of action has a negative moment (i.e., dialectical negation) where the act folds back on itself (i.e., reflection). This is why humans have always been referred to as "thinking animals." My assumption is that reflection is embedded in the forward movement of human action, which I will refer to as *agency.* I am assuming the primacy of the self as agent according to MacMurray (1957); that is, the reflective self (self as subject) is embedded in the agency of significant activity for other selves. The position being developed here, therefore, denies the possibility of a psychology based solely on the epistemological subject. The self as knower or the self as subject is not the focal point of our concern. It is here that I part company with much of the recent research on metacognition (e.g., Brown, 1978). To say this is not to deny the importance of the epistemological subject and *cognitional activity.* I am simply saying that the self as knower (epistemological subject) must be subordinated to the wider action context of human action. Several characteristics must be assumed of the *self as agent.*

CONSCIOUSNESS. The *human agent* is a conscious agent; that is, the essence of its activity is a dialectical process of *standing over against* while standing out from the *other. Other* here refers to all that is not. At the primordial prereflective level, this consciousness from the point of view (this is stretching the point) of the actor is an *I am,* since we are postulating the necessity of a human mode of intelligibility in order to interpret adequately what we are calling human action.[4] We have already hinted that this species-specific characteristic is the

---

[4]We are not saying that organic modes of intelligibility deny the phenomenon of consciousness. We are saying, however, that the self-conscious intentions of human agents are not incorporated systematically into the system of interpretation. Piaget is a case in point. Although he explores and describes conscious processes in elaborate detail, he ends up by articulating structures that are not related to the self-conscious intentions of the subjects studied. Intentions are there in many of his descriptions, and the perceptive reader can use this material with much gain for his or her own purposes.

peculiar propensity in humans for using language. Human consciousness, as one can see in the detailed observations of Piaget and in infant development, moves toward language and is significantly altered and regulated by language thereafter. The capacity for language, and the orientation toward a linguistic community for socialization into that language, makes humans a peculiar species, with a consciousness demanding a different level of intelligibility that cannot be reduced to organic adaptive processes or mechanisms.[5] This will become clearer as we elaborate the other characteristics of human action.

INTENTIONALITY. This characteristic of human action is not independent of consciousness and, in a certain sense, it is an alternate definition of consciousness that amplifies the *objective* rather than *subjective* concerns of conscious activity. To say that consciousness is consciousness of an object or other is to define the *intentionality* of consciousness. Consciousness is, therefore, not a noun but rather a verb. In short it is a process.

> The reflexive monitoring of conduct refers to the intentional or purposive character of human behavior: it emphasizes intentionality as process. Such intentionality is a routine feature of human conduct, and does not imply that actors have definite goals consciously held in mind during the course of their activities. (Giddens, 1979, p. 56)

To be is by definition to "stand out"—over against. Because we are assuming an "action self or subject," we postulate that the verb *to be* is embedded in a wider action context. In other words, the primordial sense of *existence* ensues from an active contact with the other (i.e., all that is not the self). Consciousness is the result of the multidimensional contact with the environment. In the broadest sense, consciousness is at least a property of all animals. We are using *consciousness* here as a prereflective sense; that is, it is not synonymous with intellectual activity but at the same time not totally independent of it where human actors are concerned. Merleau-Ponty (1963) clarifies the distinction we are making:

> Rather, consciousness is a network of significative intentions which are sometimes, on the contrary, lived rather than known. Such a conception will permit us to link consciousness with action by enlarging our idea of action. Human action can be reduced to vital action only if one considers the intellectual analysis by which it passes for a more ingenious means of achieving animal ends. (p. 173)

The last sentence in the quote above opens up the distinction between organic (vital) and human orders of consciousness. So far, our discussion of consciousness makes no clear distinction between organic and human modes of consciousness.[6] To make this distinction is not to deny organic biological modes.

---

[5]This collapse of the organism/environment dialectic can have serious consequences in the interpretation of phenomena. For example, McNeil (1966) can attribute an incredible power of the generative structures of language to the organism to the point where the environment, for all practical purposes, is ignored.

[6]Our position is not as drastic as Shotter's (1975), who believes that from the point of view of this discipline, it is not necessary for psychologists to have extensive knowledge of the nervous system. On the other hand, a complete knowledge of our reasons for acting is incumbent.

It assumes that animals other than humans are capable of conscious activity but also that human consciousness differs from consciousness in general in the same specifiable ways. This is a crucial distinction. If *action* is intentional, it must be directed outward to an alter-self or other. By introducing the concept of consciousness, one should not be accused of subjective idealism. Reflection or thought is the reflective mode of intentionality. It is by definition human action turned in on itself (i.e., reflect or mirror). This again is not to downgrade human reflective activity but to see it in terms of a wider intentional activity. Some of the most dramatic examples of this distinction can be seen in Piaget's observations of the infant's sensorimotor activity. These observations can give the reader a "bird's-eye view" of what we mean by intentional action if the reader does not have children of his or her own. Piaget gives astute examples of what I mean by the intentionality of reflection. (See, for example, his observations under secondary–circular reactions.) Consider one of Piaget's observations made on his child Laurent at the age of 19 months:

> Laurent, from 0;4 (19) as has been seen (Obs. 103) knows how to strike hanging objects intentionally with his hand. At 0;4 (22) he holds a stick; he does not know what to do with it and slowly passes it from hand to hand. The stick then happens to strike a toy hanging from the bassinet hood. Laurent, immediately interested by this unexpected result, keeps the stick raised in the same position, then brings it noticeably nearer to the toy. He strikes it a second time. Then he draws the stick back but moving it as little as possible as though trying to conserve the favorable position, then he brings it nearer to the toy, and so on, more and more rapidly. The dual character of this accommodation may be seen. On the one hand, the new phenomenon makes its appearance by simple fortuitous insertion in an already formed schema and hence differentiates it. But, on the other hand, the child, intentionally and systematically, applies himself to rediscovering the conditions which led him to this unexpected result. (Piaget, 1963, p. 103)

This observation is from a host of examples which exemplify a contention made by Gauld and Shotter (1977) that the infant, if observed astutely, possesses the rudiments of intention, action, conception, and communication.

INTENTION. Human action is not only intentional in the broader sense already described but also more specifically governed by *intentions*. Many of the activities of the human organism are not intended, they are *movements* or *events* in the broader biological sense (e.g., digestion, elimination, reflex activity, etc.). We can never think of synaptic transmission as a biophysical mechanism. We are calling the above *movements* or *events*, which in humans are coordinated and subordinated increasingly as the human develops to higher mental processes. In short, human intentions under normal conditions of psychosocial development increasingly come to coordinate and integrate biological adaptations. This is always a matter of degree, so we do not hold the individual or agent responsible for all his movements. For example, you do not blame a person for a knee-jerk reflex that causes injury. A human act is an *intentional* act if we refer to an *agent* as its source (MacMurray, 1957). I distinguish here human agency from a computer program in that the sources of goal-directed activities of humans are consciously regulated processes. I would call the work of a computer program an event. To call some activity an event or movement is to refer to a nonagent or process. A simple distinction that MacMurray (1957) makes here helps to make a further point. He says: "We express the distinction between acts and events; therefore, if we say: for an event there is a cause; for an act there is a reason" (p. 148). In other words, for an event or movement we ask the question "What caused it?" For an *act* we ask the question "Who did it?" "Why did he or she do it?" "What are his or her reasons for doing it?"

Harré and Secord (1972) see the role of psychology as a discipline as trying to understand the *reasons* for human action. They, in essence, reject mechanism as a mode of intelligibility for psychologists to pursue in the understanding of human action. They distinguish between causes and reasons. They maintain that reasons, which are related in some logical manner to what they explain, and causes, which are related to some physical mechanism, can appear in the explanation of happenings (events, movemnts, etc.), whether these happenings are actions of people or movements of animals. For Harré and Secord (1972) *reasons* and, in our terms, *intentions*, are usually offered in human affairs in the course of providing "justification or excuse for some action" (p. 11). In rejecting mechanism, Harré and Secord (1972) maintain that "Social behavior must be conceived of as actions mediated by *meanings* not responses caused by *stimula*" (my italics; p. 29). Therefore, in adapting the concept of action rather than movement, Harré and Secord (1972) maintain that a *person* as opposed to other species of animals does things for *reasons*. Harré and Secord (1972) project a certain uniqueness to what we are calling "human intentions." They comment:

> We share with animals the power to initiate action; that is, it is not only people who are agents, but our special powers enable us not only to monitor our performances, but because we are aware of our monitoring and have the power of speech we can provide commentaries upon and accounts of our performances and plan ahead of them as well. It is the existence of this (plans) while they are going on (commentaries) and in retrospect derivative capacity to comment upon our actions in anticipation (accounts), that is our most characteristic feature, and around which the science of psychology must turn. (pp. 90–91)

Harré and Secord (1972) refer to the unique character of human intentions as "teleological." They, with other theorists within the more recent tradition of language analysis, make a crucial distinction between causal and teleological explanation (see Bernstein, 1971; Gauld & Shotter, 1977; Harré & Secord, 1972). We feel that this distinction is not sufficiently subtle and serves to obliterate the important distinction between the "organic" and "personal" modes of intelligibility. It is true that psychological inquiry conducted at the mechanical mode of intelligibility eschewed teleological explanation. It is also true that genuine organic interpretations of events easily adopt teleological mechanisms as part of their explanatory apparatus. For example, a species-specific goal-directed activity such as imprinting can and should be called a teleological act. To leave human intentions at the level of organic intelligibility (a teleological explanation) is, in our opinion, deficient. If an *event* (i.e., mechanical mode of interpretation) is conceived of as different from an *act* (i.e., personal mode of interpretation), we would also wish to say that human intentions or plans are not teleological processes. In the stricter sense, one understands the behavior of organisms teleologically in terms of adaptation (MacMurray, 1957). In the above sense, we do not assign a *reason* but rather a motive force. For example, certain animal tropisms (nesting, flock movement, the "dance of the bees," etc.) which can clearly be interpreted as teleological mechanisms (i.e., goal-directed and adaptive) cannot, at the same time, be considered an *intention*. We feel that it is necessary to say with MacMurray (1957) that human action as distinguished from other animal action is not teleological but *intentional* (see also Taylor, 1964). For us, then, the human action based on human intentions provides a crucial distinction between the organic and personal modes of intelligibility. The almost infinite adaptability of human beings makes action based on human intentions an *open-ended process* rather than a teleological adaptation. By *open-ended*, I mean that human action has a "projectual" character that goes beyond adaptation into construction. This constructive aspect will be elaborated in greater detail in the next chapter, where I introduce the notion of "project." Human action is not fully determinate. The end or goal of an intention cannot be specified because human action is to a very great extent limited and supported by the "other" (i.e., resistance). Since human intentions must be coordinated with the intentions of other selves, intentions must be modified in the process of interpersonal interaction. Where this is not possible, one refers to a person as "rigid," stylized, overdetermined, or, in my case, bullheaded. So, the essence of human intentions is their relational and open-ended quality. This relational quality of human intentions, which we will have further occasion to discuss, is summarized by Shotter (1975) when he says:

> The person's own self and the selves of others are both then discovered together in relation to one another; they are reciprocally determined within the same categories as one another. In other words, to be me I need you: I need you to respond to my movements for me to appreciate that my movements have consequences in you. I need you to respond to the meaning in my action for me to appreciate that my action does in fact have a meaning. (p. 110)

Shotter's (1975) position is helpful for now in hinting at the relational quality of human action. We will have to expand it in order to deal with larger structural totalities such as class and gender when we deal with these concepts in Chapter 4. For now, let us provisionally treat human intentions as coordinated and open-ended acts. To speak of them as teleological is to restrict their human quality, which demands a personal mode of interpretation.

This brings me to a final point before we move on. I would agree with others (e.g., Harré & Secord, 1972) that human intentions are the essential ingredients for psychological inquiry. This does not mean that humans cannot be studied under more general characteristics (i.e., mechanical or organic modes). Our position is that this is not psychological inquiry. For us, psychological inquiry involves a movement into the "personal mode of interpretation," tackling human intentions head on, rather than as derivatives from the mechanical and organic orders.

RESPONSIBLE. To speak of *human action* as we are developing this term is to demand a discussion of an ancillary concept, responsible action. Everything I will say on this concept has illustrious forebears. (Book III of Aristotle's *Nicomachean Ethics* is a case in point.) Responsibility is one of the characteristics of a human act insofar as we are accounting for the actions of a person in terms of their desires, intentions, and purpose. This in humans, of necessity, involves the negative pole of action; that is, reflection. Responsible action, has therefore, as part of its ongoing process, a reflective component which in ordinary language we call "deliberation." The attribution of responsibility for an action is canceled out altogether if it can be shown that the behavior concerned was not governed, at least in some of its aspects, by intentions (Taylor, 1964). The deliberation involved in the reflective component of human action raises the thorny issue of the autonomy and freedom of human action.

> Being able to deliberate before one acts (or in the course of a complex action), and as a result make clear to oneself one's own reason for action, is part of what it is to be an autonomous responsible person, not reliant like a child upon others to complete and give meaning to one's acts. (Shotter, 1975, p. 96)

I understand, as do most western systems of jurisprudence, that part of the capacity for responsible action emerges in the psychosocial development. This is not a foregone conclusion, however, since we find extreme cases of adult individuals who, for all intents and purposes, are irresponsible (i.e., lack the capacity for deliberation). One talks sometimes of individuals being creatures of their surroundings. Another way of talking about the lack of responsiblity is to say that a person is ruled by his desires. Desires, here, are seen as potent motives that lack a component of reflection or deliberation. The notion of a deliberate action assumes, after the process of deliberation, that there is a motive force for that process which has consequences that are unique to that process. In short, a responsible act must, in some sense, make a difference in some outcome. This raises the question of whether responsible action is free. Chein (1972) maintains that the notion of freedom rests on three premises: (1) that there are such things as volitional desires, motivations, and so on; (2) that volitions have behavioral consequences; and (3) that volitions are not reducible to variables of the physical environment or to variables of physiological processes. I am using the term *intentions* in the same way that he uses *volitions*. I would add one more premise to Chein's in order to call a free act responsible, and that is that volitions must be capable of undergoing reflective scrutiny (deliberation). To say that a person is *free* is to say that motives or intentions are a constitutive part of human action. Chein (1972) would say that conditions for an action are, therefore, of three kinds: (1) constitutional, (2) environmental, and (3) motivational. His understanding of freedom, from which we draw, should not be understood only in these schematic terms. There is a great deal of subtlety in the crevices and interrelations of these components. The subtlety is seen in the following reflection:

> Behavior is free to the extent that the environment and the consitution do not dictate or preclude a particular action, the greater the degree of freedom that exists within the limits of the imposed constraints. Within the limits of these constraints the behavior that actually occurs is determined by the motivation of the actor. No behavior is ever completely free, some degree of freedom being lost as a consequence of the constraints of constitution and environment but degrees of freedom are also gained from the dependabilities of constitution and environment. The gains and losses, however, do not cancel each other out because they are different in kind. In addition, with respect to any particular behavior, some degree of freedom is lost to the constraints imposed by competing behaviors. A person may be free to the maximum degree that environment and constitution permit if and to the extent that, he does not lose degrees of freedom from constraints imposed by motivational conflicts, sacrifices relatively trivial motives, and effectively sequences or integrates the others. (Chein, 1972, pp. 33–34)

To talk of the free, responsible action must, therefore, be seen within a structure of freedom.[7] We will later consider this as the "intentional intervention of the project." For now, I am not talking about some isolated act free of resistance. The structure of freedom is an agon. The free human action, however graceful it may appear, is the outcome of a tension between all of the factors that Chein (1972) elucidates. This agon or tension, as we are calling it, should not be seen simply in a pejorative light. A person is said to have character when her or his actions are said to be reflectively motivated. This is a complex blend of motivational factors, environment, and constitution. A person is said to be responsible when it can be assumed that his or her actions are under the control of his or her motivational powers. To be a responsible agent is to have personal powers that make a difference to the outcome of actions. This brings up an important distinction which can be called the "agent–patient dialectic" (see Harré & Secord, 1972). The concept of the responsible agent is a one-sided understanding of human action if we do not acknowledge that the agent not only acts but also undergoes or suffers action that is partly out of her or his control.

---

[7] We are raising the question of a free act in this chapter from an egocentric point of view; that is, from the point of view of the agent. As our treatment develops in the subsequent chapters, we will be able to talk about a relational freedom or "our freedom."

We call this "patienthood." To deny the phenomenon of patienthood in human experience would fly in the face of much of human experience that we all have access to in our personal world. This "patient experience" is not only a result of external (i.e., environmental) factors, it may also be a result of our constitutional makeup as competing motivational factors that impede the possibility of human action. We also become slaves to our own habits. The free, responsible act must be seen within this dialectic of agency and patienthood. This dialectic is quite complex and subtle, as we know from our own personal experience. To think of ourselves as only agents is pretentious. To think of ourselves as patients is to attempt to excuse all our actions as beyond our personal powers. Responsible freedom is a limited freedom worked out within the dialectic of agency–patienthood. The prayer, "God grant me the serenity to accept the things I cannot change, courage to change the things I can, and the wisdom to know the difference" expresses this complex balance. To talk of "human freedom" and "responsible action," though, in the end assumes that the self as agent incorporates its own patienthood (i.e., self as subject). A critical ethical assumption pervades this book, and it should not go unacknowledged; that is, that individual persons and the wider community proper have the responsibility of creating institutions that enhance the use of personal powers. I call this, in the next chapter, the "human project." At the individual personal level, to relinquish the possibility of using one's personal powers is irresponsible action—you might say, "playing the patient." At the larger social level, the creation of societal institutions that rob humans of their personal powers, that is, their capacity for responsible action, is by definition an alienating society: I–we share the responsibility to transcend situations that help to create alienated individuals (i.e., patients) and alienating societies. This is a complex issue and a dialectic itself (i.e., alienated individuals create alienating socieities, and vice versa). We will develop this further in subsequent chapters.

SIGNIFICANT. At first blush, it would seem like rhetorical flourish to say that human action is significant. To characterize human action as significant within the perspective we are developing is to emphasize its *sign* or linguistic quality. We will move on in the next chapter to develop the notion of human actions as expressions of a personal world. To say here that human action is significant is simply to anticipate, from the point of view of the actor, that his or her actions are expressions for others. To say that a human act has significance is to say that it has a meaning for others. It is not always easy for agents to find an audience for their expressions, but in the end there must be some audience, however sparse, to make the judgment of the significance of a human act. There is no such thing as a significant act in itself. The meaning or significance of a human act is the place that it occupies in a network of *relationships* (Chein, 1972). Therefore, there is no such a thing outside a relational context. This is not to say that human action is not misunderstood or ignored for either benign or malicious reasons. We will later be critical of social structures that systematically ignore the significant action of certain groups of actors. Under these conditions, these social institutions rob particular agents or groups of agents of their personal powers. Part of the task of a critical psychology will be to elucidate and condemn these types of social structures and demonstrate their deleterious effects on human agency. To do this, we must move out of the egocentric movement we have just developed and consider human action in its wider relational context under the term *personal world*. Before doing this, let me briefly tie up some loose ends and summarize where we have come thus far with this chapter.

Loose Ends: Habits and the Unconscious

Habits and Intentions

This chapter has focused on the intentional base of human action from the point of view of an agent. In this light, we are all creatures of human intentions. Our western legal structure assumes this in attributing responsibility for human actions based on our intentions. A crime has to be *intended* to warrant culpability.

At the same time, our common sense tells us that we are creatures of *habit*. We do many things in a consistent pattern, which is not governed by intentions. Yet, we do not call these *reflexes* because they were acquired through learning. Some of these habits have been learned by unconscious associations, others through our intentions to acquire habits. The whole process of intending to acquire habits we call *discipline*. What must be seen here is that habits and intentions are not separate from one another; rather, they are dialectically related. It is, as we shall see later, the dialectic of freedom and determinism. Much of what constitutes psychology in this century is a solar emphasis on determinism. For example, all behavioral learning theories have made *habit* a core concept. We are now seeing a renaissance of the intentional nature of humans in the recent emphasis on metacognition (Brown, 1978). Here, however, human intentions are seen in a vacuum from the habitual. What is beneficial in the recent emphasis on metacognition is the regulative effect of intentions on human actions. This emphasis is laudable. Intentions can create habits and can transform those that have been learned through unconscious association and are no longer functional. The latter function is what I understand by the process called "working through" in Freudian psychology.

The problem here is that all points of view in this chapter concerned with intentional action are partial and egocentric. By egocentric, I mean from the viewpoint of the single agent. I intend to introduce a corrective in the next chapter, where habits and intentions will be broadened by what I will be calling the *habitus* and *project* of the personal world. In other words, I will attempt to bring the notions of intentions and habits to a broader cultural and sociological horizon. This corrective will be seen when I critique the notion of individualism in the next chapter.

Intentions and the Unconscious

Habitual actions are unintended. This does not mean that the original action that went into making the habit was not intended. Habits do indicate that a pattern of behavior is no longer governed by intentions. In this sense, one can speak of the state of a habit as passive. There are other unconscious actions that do not fall into the category of habit. I am referring to a process, illuminated by Freud, called "repressed intentions." Freud called repressed intentions the "unconscious." For Freud, the unconscious is an *active, dynamic* process whose intention is to *deny* the presence of unacceptable intentions (i.e., the dynamics of superego). Freud has stood the test of time, and it would be difficult to bypass this Freudian contribution. In the Freudian paradigm, the mechanisms of the unconscious operate within the microcosmic world of family dynamics. Psychologists of the psychodynamic variety believe that this is their territorial imperative and boundary. Marx introduced a similar concept from a more macrosocial level by introducing the notion of false consciousness. For our purposes, false consciousness is a set of repressed intentions about the true nature of societal dynamics. Specifically, Marx located the repression in the denial of the economic motive, which pervades capitalism. Many modern social theorists would contend that these two different repressions should be seen dialectically. What is suggested is a merger of Marx and Freud (see Becker, 1975). I do not intend to engage this literature, but my treatment will take into account its concerns. What is at issue when speaking of repressed intentions is the presence of "unfreedom." A critical psychology attempts to open up repressive mechanisms in the micro- and macrocosms and will therefore challenge the very nature of these repressions while advancing the cause of freedom. At this point, my statements are simply rhetoric. We must put flesh on these bones.

SUMMARY

We have developed the notion that human beings are actors whose actions are communicative expressions. This brought home the notion that human action is communication, which makes it a phenomenon of language broadly conceived. Expressions are made to be interpreted and, therefore, the interpretation of expressions is a desire to make them intelligible (i.e., meaningful). Psychology as a profession has no exclusive role in the interpretation of human expressions, although we will argue later for the possibility of a unique role. Nevertheless, we have argued in this chapter that psychology in all its manifestations is in the business of the interpretation of human expressions. From there, we go into the adequacy of the interpretative system being utilized in the interpretation. We have distinguished three broad systems of interpretation and called them modes (i.e., mechanical, organic, and personal). Each of these modes has broad characteristic features that allow us to distinguish them from one another and also to locate them in the history of psychological research and theory. It is obvious from our treatment that we do not treat them as equal in merit. This book will emphasize the mode of the personal as we have developed it in this chapter. All these modes of interpretation are sensitive in their own peculiar ways to the problems that language presents. The mechanical mode of interpretation systematically tries to deal with ambiguity of interpretation by eliminating the ambiguity with operational definitions. We question whether any system of intelligibility can go beyond interpretation (i.e., totally eliminate ambiguity). Mechanism as a mode of interpretation frequently used the machine as an analogue for the understanding of human expressions. This mode also rests on the assumption that humans are physical systems, just like all other systems in nature, and can be best understood in terms of causal mechanisms. There is no apparent need in this system to conceive of human beings as *actors*.

To speak of an organic mode of intelligibility or system of interpretation is to move from physical to biological systems. The organic metaphor acknowledges that *living biological systems* have their own unique status in nature and need to be systematically interpreted in this light. Humans are seen as one organic system in an adaptive evolutionary process. To understand a biological system is to come to terms with the structure and functions of that system in nature. Organic modes in all their variety are holistic in nature and acknowledge that human complexity cannot be simply broken down (analysis) for purposes of understanding. Thus the system of interpretation is usually constructive. The human being as a biological system is seen as an *active system* that transforms and is transformed by its environment. With all their complexity, organic systems of interpretation do not deem it essential to understand the phenomena of consciousness and humans as conscious agents *per se*. This does not necessarily mean in all cases that consciousness and its role is denied. It does mean, however, that it does not, in most cases, play a prominent role. In Chapter 2, our rudimentary presentation of a personal world focused on the unique characteristics of human activity, implying that it is necessary to go beyond the mechanical and organic modes in order to characterize human expression. Our system of interpretation or mode of intelligibility for a personal order is premised on the unique characteristics of the personal world which, in its most basic way, is seen as an interrelation of human actions involved in communicative acts. There, we elucidated that a human act has characteristic features: consciousness, intentionality, intention, responsibility, and significance. We must now see these features in their interpersonal context in the linguistic world we are calling the "personal."

3

THE PERSONAL WORLD
A Relational Event, a Cultural Reality

The title of this chapter is rather wordy. Nevertheless, it does try to say what the work of the chapter will be. The conclusion will be that the personal world is a cultural form. To arrive there, we must take a winding road. Let me help at the outset with some guideposts in the form of an outline.

The last chapter developed the notion of human agency as expression in an egocentric manner—that is, from the point of view of an individual agent. We must now complete this expression by embedding it in the larger structural totality of a personal world (i.e., the world of the you-and-I). To do this, we must clear away an impediment; that of the phenomenon of individualism. Traditional psychology operates from the viewpoint of individualism and therefore is biased toward egocentricism (the solar I) in its interpretation of human action. I will try to show how the egocentric viewpoint is reflective of what I would call a "liberal" sociopolitical synthesis. In addition, an example and set of contrasts are discussed to show how the egocentric viewpoint works out in psychological research. After this ground is cleared, we will go on to explicate the framework of a personal world and a dialectical framework that characterizes the personal world as order, change, and meaning. Finally, we will go on to consider how the personal world as *meaning* is, in essence, a cultural form.

INDIVIDUALISM: AN IMPEDIMENT

Our conception of a personal world must now be considered distinct from a conception of the individual. In fact, the whole historical thrust of liberal individualism makes it difficult to consider the relational totality of a personal world. Some important distinctions and the history behind them must now be made, for reasons of clarity and exposition. This is done to clearly distinguish a personal world from the liberal conception of the individual. To talk about a term is at

51

once to talk of its history. The term *individual* is a case in point. Williams (1976b) in his *Key Words* gives a capsule history, reaching back to medieval times, of this term. In the medieval world, the term *individual* meant "inseparable," and the term was used mainly in theological argument about the Holy Trinity. In premodern times, the term was always used as an identity differentiation within a larger structural totality (e.g., one God and three divine persons). For Williams (1976b), there is a crucial change in emphasis that enabled us to think of the individual as a kind of absolute, divorcing it from any other group membership beyond itself. Subsequent social thinking has depended heavily on the latter definition of the individual as an absolute unto itself. I would call this change in emphasis "egocentric" where the ground is shifted from a you-and-I to an I. The sociopolitical consensus around this term in its modern definition is almost a counterposition of our definition of the personal world as relational. As Williams points out:

> In England, from Hobbes to the Utilitarians, a variety of systems share a common starting point of psychology, ethics, and politics. It is rare, in this tradition, to start from the fact that man is born into relationships. The abstraction of the bare human being, as a separate substance, is ordinarily taken for granted. (p. 77)

The modern definition of the individual as an autonomous social unit is a product of a consensus achieved by liberal social thought. In liberal social theory, individuals are seen as separate, autonomous monads that are unique unto themselves. The primary position of the "state of nature" is characterized by Hobbes as solitary, where the individual contracts ("social contract") out of fear for survival. The creation of a world (i.e., society) is based on a contractual arrangement of separate individual entities. Liberal ideology refers to a style of thought developed at the time of the French Revolution (Mannheim, 1953a). Three pervasive ideas characterize the synthesis of "classical liberalism" (De Lone, 1979). The first idea is that the universe is governed by natural rational laws. This is the core belief for the integrity of natural scientific inquiry. Second, that this natural law could be deciphered or interpreted by the operations of human reason. Finally, the individual's well-being is the ultimate point of societal synthesis. This represents a revolutionary departure from the medieval synthesis—where the stress was on community synthesis while individual integrity was regarded as heresy. The preliberal synthesis was communitarian in nature but with a blunting of the individual for the purposes of community consensus. Therefore, liberalism marks the accent of the individual over and against the community.

The development of thought on natural law at the time of the French Revolution was truly a revolutionary event, which marks the ascent of what is today called "individualism." A central characteristic of the liberal synthesis is the claim, at an abstract level, of universal validity for every individual (Mannheim, 1953a). This was achieved through a metaphor of atomism and mechanism; collecting units into larger totalities additively (e.g., the state, etc.) (Mannheim, 1953a). Here, it is important to point out the close relationship between liberal social and political theory and the historical development of psychological theory. The interweaving of psychology and social theory generated an *atomistic* view of political economy. The conception of the individual as separate (i.e., a social atom) fed into the atomistic view of political economy, society becoming the sum of individual actions and decisions (De Lone, 1979). Here, the psychological tradition of Locke combines with the economic and social theory of Adam Smith, suggesting that psychological attitudes (i.e., abilities, personality, moral traits, etc.) prepare individuals to become signatories to the social contract (De Lone, 1979). Rationality, in the liberal tradition, will be allied to autonomy, reason being the compensating mechanism against unruly desires that threaten the viability of contractual relationships. In the liberal tradition, *reason* is expressed as the universal, formal, abstract, thoughtful, and so on, in contrast to the irrational, which is particularistic, content-laden, concrete, and emotional (Unger, 1976).

The wisdom of hindsight allows us to lay bare some of the distortions of the conception of the human as a conglomeration of autonomous social actors. A more communitarian perspective takes issue with the liberal contractarian view by interjecting a historical perspective. A radical communitarian formulation assumes that a conception of society should be sensitized to historical occurrences (Sullivan, 1977a, 1977b). The person within a communitarian conception is embedded in real historical relations. She or he comes into a world that is already a momentum and where there is a solid, weighty, and dense social structure in which the person is influenced and which he or she operates upon. The personal world I am developing here is embedded in larger structural totalities that are impersonal in nature but nevertheless affect the viability of the personal world. The world of monopoly capitalism and concentrated economic conglomerates (e.g., multinationals) seems remote from what psychologists would call a personal world. This type of isolating is essentially the basic problem to which atomistic theorizing falls prey. Psychologists study individuals and groups as if they were independent of larger social structures. Economists proceed to theorize as if economics were independent from the personal psychological world. When a major multinational relocates to another region for economic gain, no one holds this type of conglomerate responsible for the enormous personal costs on community lives which are affected by sudden and possibly chronic unemployment. Nevertheless, we know that high sudden unemployment is correlated with deviance, alcoholism, depression, and suicide. This brief excursion is intended simply to illustrate a point that I will develop more fully in the next chapter: that an adequate treatment of a personal world cannot be simply psychologistic in nature. The personal world must be embedded in social and political conditions which *situate* the interpretation.

The Impediment

The possibilities for developing this line of social analysis have historical realities which are weighted against it; we now turn to these. Psychology as a discipline is one of the embodiments of individualism. To conceive of a discipline that could carve out for itself the study of the microcosm called the individual is truly a western historical creation. Disciplinary boundaries are in actuality permeable, but they nevertheless express some clear divisions of labor among social inquiries. Psychology, as a discipline, represents the overarching belief that the study of individual and microsocial units is possible and desirable. To be able to demarcate sociology from psychology represents the belief that the individual can be studied as relatively independent from a wider social setting (i.e., social structure). A distinct discipline called psychology is truly a symbol of the liberal tradition, since it is the professional embodiment of the individual as a totality or unit. Psychology as a discipline is truly an expression of liberal individualism.

The fragmentation produced by rigid disciplinary boundaries has had consequences, as to how we are able to think about the person. The person as individual is a conception of the person conceived of as independent from social and historical conditions. Sève (1978) delineates some of the consequences of this type of fragmentation:

> One may at first detach the personality from the social conditions in which it is formed but in doing so one deprives oneself of any way of accounting for its deep-seated *sociality*, and encloses oneself in a hopelessly abstract, non-historical conception of individuality whether in the form of a spiritualism of the person or a biologism of temperament. In either case, the essential historicity of personality escapes. (p. 31)

Riegel (1978) conjectures some of the reasons why a dialectical conception of personality has eluded the discipline of psychology systematically. In his review of diverse attempts to formulate personality theories, he notes that these theories have by and large (despite their differences) been theories which have primarily dealt with individual psychological conditions. For the most part, cultural–sociological considerations entered the discussion only secondarily; presumably they are the task of the sociologist. Lewin broke some new ground by proposing a "field theory" of personality, but it was clear that his formulation was safely within the horizon of individualism. His emphasis on the unique experience of the individual in unique situations committed him in the direction of an individualistic perspective (Riegel, 1978). Riegel (1978), in reviewing this situation, contends that it is only with the development of social and psychological theories (e.g., George Herbert Mead), which concerned themselves with social roles and the social self, that we see a preparation for an alternative away from traditional personality theory, which failed to consider the codetermination of self and the other in the developmental historical context.

The Personal World as a Relational Totality

The viewpoint we are developing is that the person is person only as I-and-you. We are clearly departing from the liberal tradition (see also Sampson, 1977, 1981; Unger, 1975, 1976). There is, therefore, the clear acknowledgement of the existence of collective subjects, which clearly violates the principle of individualism as exemplified in the liberal tradition (Unger, 1975). At a later point, we will refer to this notion of collective subjects as culture. Initially, I would like to explore several reasons why it is desirable to shift to the viewpoint of the I-and-you as suggested by MacMurray (1961). In doing this, one must shift the ground by acknowledging the existence of collective subjects in violation of the principle of individualism. One must countenance the notion of a person as a reciprocal relation. A person is at once subject and object, encompassing both modes simultaneously. The personal world is a unity (I–thou); the person as *subject* is an I and an *object* is a you, since the you is always the other (MacMurray, 1961). Historically, the child development literature has assumed that the child's interaction with objects (nature or I–it relationships) is an adequate analogue for understanding interpersonal interaction (I-you) (see Richards, 1974). A conception of the "unit of the personal" assumes that there is a special class of objects from which the conscious self must also be separate; it is the class of objects who are also subjects (i.e., other selves; Unger, 1975). The child development literature is now developing a corpus of research which indicates that even newborn infants employ different modes of response when relating to people as opposed to things (see Lock, 1978; Richards, 1974). For instance, the vocal responses of 3-month-olds can be much more effectively conditioned with a social than a nonsocial reinforcement (see Richards, 1974). As early as 3 weeks, the young baby expects the human voice to appear in the same orientation as the human face (Ausubel *et al.*, 1980).

Much of the earlier literature on early child development neglects the effective environment in which the child develops; no description is given to the reciprocal interaction of the child with other persons (Ausubel *et al.*, 1980; Ryan, 1974). There is no characterization of the speech of others to the child, and thus the child's speech is treated as an isolated phenomenon (Ryan, 1974). This is illustrative of the individualistic orientation we have just dealt with under liberalism. There is considerable evidence, in contrast, that very early babbling is not random and reflects features of the language spoken around the child. For example, as early as 4 to 6 months, the babbling of infants could be distinguished between Chinese and American parents. Systematic selectivity can be seen in the very production of sounds before speech appears (Ausubel *et al.*, 1980). Ryan (1974) makes the point from her own research that any analysis of human communication must include a description of the structure of intersubjectivity between participants in a dialogue (i.e. I–you–you–I, etc.). In following the lines of analysis as suggested above, Trevarthen (1980) concludes from his studies on infants that within the first few months of life the caretaking interaction indicates powerful evidence of new forms of "joint intentionality." The child, from birth, must be seen as an active participant (agent) in a jointly reciprocal relationship with his or her caretakers (i.e., I–them), rather than a passive object of socialization practices. Communication expresses the cognitive aspects of this reciprocity. The development of individuation within this larger totality is what one calls consciousness. Consciousness is the experience of separation and limitation (i.e., over and against others and nature; Unger, 1975). Consciousness as we are defining it must be seen in a relational totality.

> Consciousness implies autonomous identity, the experience of division from other objects and from other selves. But the medium through which consciousness expresses itself is made up of the symbols of culture, and these, according to the principle of totality, are irreducibly social. When you speak of language or make a gesture, you perceive and communicate meaning in categories that are the common patrimony of many men. By what power can you and they speak to one another? It must be possible for each to view the other's statements and acts as the signs of certain intentions. These intentions can, in turn, be understood because they are intentions you too might have. It follows that consciousness always presupposes the possibility of viewing other persons as selves that could, under favourable enough circumstances, see what one sees and believe what one believes. That is the cognitive aspect of sociability. (Unger, 1975, p. 215)

The cognitive aspect of sociability assumes that some of the above features must be present as early as infancy in order for the child to build up communicative skill (Ausubel *et al.*, 1980).

There is also a moral aspect to sociability, and it would appear to be a requirement for the persons that he or she be recognized by others as a person (Unger, 1975). The absence of this aspect of sociability can have most deleterious effects on the development of the person. The importance of this relational reciprocity cannot be underestimated and ignored. Ausubel *et al.* (1980) indicate that if these schemes are not securely built, either through misfortune or mismanagement or some endogenous defect in the developing child, the whole system of social skills and affective understandings is put in jeopardy. The constant message in the new work on caretaker–child interaction is the centrality of the interpersonal relationship in the transmission of meaning.

It would seem that humans are possessed of a compulsion to share their conscious understanding and wants as intimately as possible (i.e., dialogue; Trevarthen, 1980). When this dialogical relationship is fragmented or broken, as seen in maternal deprivation, serious consequences ensue (Ausubel & Sullivan, 1970; Ausubel *et al.*, 1980). The early studies on maternal deprivation report findings indicating that after 6 to 9 months of age, prolonged or severe deprivation of maternal care or abrupt separation from the accustomed mother figure during the first 3 years of life often (but not necessarily) leads to extremely serious developmental consequences. Although there are reservations and subtleties to this picture (see Ausubel & Sullivan, 1970), I quote them here to score the importance of the necessity of the relational components (i.e., intersubjective and interpersonal) of the personal world from its origins in infancy. This literature, from both its positive and negative aspects, makes it impossible to think of individuals in terms of isolated units (individualism). The core unit is relational and must be characterized as a *dialogue* (I–thou). From our analysis, it can be stated that the fundamental motives for developing cultural forms are already manifest in prevocative form in infancy (Trevarthen, 1980). From these early infant developments an elaborate sequence of progressively more powerful stages of individual motives and awareness ensues (Trevarthen, 1980).

It is not my task to develop the child-rearing literature in a relational form. Our exploration into this literature was to make the point that the personal world as a relational totality is a given rather than an achievement of human culture and a total necessity or prerequisite for human development.

THE PERSONAL WORLD AS A METAPHOR OF COMMUNICATION

In the first chapter we hinted at the metaphor of the personal world as being that of the communication process itself, which assumes the primacy of the notion of human action as expression. With the assumption of language as fundamental to human existence, it follows that the human sphere (i.e, the field of the personal) cannot be understood adequately through organic or mechanical categories (MacMurray, 1961). The core unit of a communication is dialogical in nature. Human expression, as we have outlined in the previous chapter, must be understood in a dialogical context. In other words, the unit of analysis is not an I but an I-and-thou. Winter (1966) outlines the core features of a dialogical model in Table 1. His model presupposes two tensions: (1) the tension of a *self* coming to sociality through gesture, sign, and symbol or what we are calling "human expression" and (2) the balance of the gesturing self (dynamic) with the meaning-receiving and meaning-giving interpretative response of others. Table 1 identifies three basic elements which he draws from the pioneering work of George Herbert Mead. These elements are (1) dynamic, (2) form, and (3) unification. The *dynamics of a gesture* is an attempt at eliciting attention, the gesture being seen as standing one against the other and reaching toward the other. This is one element of a polar tension. The *formative* element is an interpretative response and constitutes the other side of the polar tension. The interpretative response as *form* is a meaning-giving act (i.e., form receiving as in listening and form giving as in speaking). Unification is the crucial element in this analysis (Winter, 1966). The preeminence of *unification* becomes manifest in the reciprocity of perspectives, which is presupposed in seeing the gesture from

TABLE 1.   A Dialogical Model

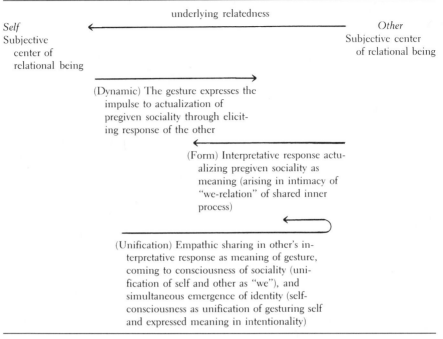

underlying relatedness

Self   *Other*
Subjective   Subjective center
   center of   of relational being
   relational being

(Dynamic) The gesture expresses the
impulse to actualization of
pregiven sociality through eliciting response of the other

(Form) Interpretative response actu-
alizing pregiven sociality as
meaning (arising in intimacy of
"we-relation" of shared inner
process)

(Unification) Empathic sharing in other's interpretative response as meaning of gesture,
coming to consciousness of sociality (unification of self and other as "we"), and
simultaneous emergence of identity (self-
consciousness as unification of gesturing self
and expressed meaning in intentionality)

Note. From *Elements for a Social Ethic* by Gibson Winter. Copyright 1966 by Gibson Winter. Reprinted by
permission of the author and the publisher.

the other's perspective; at the same time, this empathic seeing from the stance of
the other involves a distancing from onself which is accomplished by the mediation of the other's giving of form to the gesture. This makes it possible to share in
a common world of form. Later we will consider it cultural form. For now, the
unification of the polar tensions of *dynamics* and *form* reveals some fundamental
and essential human prerequisites. As Winter (1966) put it:

> The unity of self and other is actualized as a social world through the sharing of
> gestures, signs and language; the unity of the self is actualized simultaneously, as
> personal identity of the intentionality of the self with the responses of the social world
> and particularly of the world of significant others. The structure of dynamics and form
> in their unity as self-consciousness reveals man's essential tendencies as 1) care for
> response 2) reflective awareness and openess to form 3) thrust to integrity and unity.
> (Winter, 1966, pp. 102–103)

This "thrust to integrity and unity" (i.e., unification) should be understood as an
ontological (i.e., essential) and ethical (i.e., oughtness) category. In other words,

it should be seen as an essential characteristic for the achievement of humanness and the solidification of personhood (i.e., individuation) in a personal world (i.e., unification).

Winter (1966) contends that the problem of a human science is to be able to formulate the parameters of the intersubjective world in such a way that the self-transcendence of the intentional consciousness (i.e., dynamic polarity) is held in tension with the structures of the social and cultural world. By and large, psychology as a human science has collapsed this polarity in the historical development of psychological theories. In a very gross way, one could say that theories of psychology have been either *voluntaristic–intentional*, expressing some form of dynamics of an individual self, or conformist, expressing some form of dynamics of a social self. The theory of Piaget, Maslow, and Chomsky would collapse the tension to the side of a *dynamic intentional* self with active personal, cognitive, and linguistic structures. Behavioristic social learning theories in some very essential ways collapse and negate the importance of an active intentional self in order to stress the importance of social conformity (conformist perspectives). The former theories seem to express a form of limitless freedom because of the activities of the intentional organism; the latter seem to stress determinism as a result of being molded (i.e., conformed) to external social structures.

The Pioneering Work of G. H. Mead

George Herbert Mead (1964) would have to be considered a pioneer in the development of a social–relational view of self in his "symbolic interaction" theory of the development of the self. He was able to contrast the difference between an individualistic and a social theory of the self. From the perspective that we are attempting to develop, one would consider his social viewpoint as organic in nature. Nevertheless he, early on, was able to discuss the difference between a type of psychology which assumes a social process or social order as a logical and biological prerequisite for the appearance of selves and a psychology where individual selves are presupposed or prior to social processes (i.e., individualistic). Mead (1964) identified these different perspectives with different historical views of the state:

> The difference between the social and the individual theories of the development of mind, self, and the social processes of experience or behavior is analogous to the difference between evolutionary and the contract theories of the state as held in the past by both rationalists and empiricists. The latter theory takes individuals and their individual experiencing—individual minds and selves—as logically prior to the social process in which they are involved, and explains the existence of that social process in terms of them; whereas the former takes the social process of experience as behaviour as logically prior to the individuals and their individual experiencing which are involved in it, and explains their existence in terms of that social process. (Mead, 1964, p. 242)

Mead (1964), by taking the position of a social self, contended that the mind, self, and so on could never find expression or even come into existence except in terms of a social environment and that an organized set or pattern of social relations and interactions is necessarily presupposed by it and involved in its nature. For Mead, the personal world was a *relational totality*. He perceived personality as an internalization of roles and these roles as patterns of attitudes and actions. He recognized the indetermination of inner and outer conditions by the contrastive comparison of the I and the me's. The I represented the inner biological core of the organism, whereas the me's are achieved by the perceptions and enactments of social roles. Mead never adequately developed a clear conception of the I, and it appears as a creative but mysterious inner force. The major thrust of his work was in the development of the person through role enactments (me's). Mead's conception was a symbolic behaviorism and his work was a daring effort to anchor personal consciousness itself in the social process (Gerth & Mills, 1964). His notion of a "generalized other" (me) consolidated the possibility of linking the private and the public—that is, the innermost acts of the individual with the widest kinds of social–historical phenomena (Gerth & Mills, 1964). Mead's conception of a relational totality essentially accounts for the individual's adoption and confomity to social roles (i.e., social formations). His inability to formulate a clear conception for the I leaves his overall view most incomplete. His "generalized other" accounts for social determinisms (i.e., the reproduction of the social world in new individuals), but there is very little in his position that accounts for social change (i.e., social transformations). His conception of the I as a creative, organic entity hints at the possibility of new creations (i.e., transformation in social roles), but this element of his theoretical formulation was left undeveloped in his thinking. I will return to this issue later in another context, where it will elucidate a conception I wish to develop on human freedom. Despite the shortcomings of Mead's perspective, he nevertheless laid the groundwork for the view of the person as a relational totality with other social selves. I intend to use his core concepts to develop integrative components of a personal world.

Integrative Components of a Personal World

The accent at this point is on *personal* and *world*. A concept of the person is always related to a world. At this point it is necessary to say that the world is all that is not I. It is that resistance or counterpoint to the person. When that world constitutes other I's (persons), they are from one person's point of view "you" or "they." Obviously this relationship is reciprocal. We have already indicated the impossibility of considering persons as separate from a world. Thus, to talk about a person is to talk about a personal world. *World* emphasizes the situated character of intentional action. Thus, human action (discussed in the previous chapter as conscious, intentional, intending, responsible, and significant) must be clearly seen as existing in a world (i.e., *we* relation) if action is going to be considered communicative (i.e., expressive). Recall from the first chapter that the argument is to construe the personal world in a metaphor of communication in distinction to the metaphors of organism and mechanism.

At the beginning of this chapter (see Winter's diagram in Table 1), we have talked about the integrative components of a personal world as dynamics—form and unification. What I would like to do now is to hint at how this unification takes place within the context of a schematic model of personality integration. I will express this integration in terms of polarities which must be in tension with one another rather than as discrete categories. My treatment here is dynamic and process-oriented, so readers will be deluding themselves if these polarities are reified into static categories. Let me now outline in Table 2 the polarities as dialectical components of the personal world.

Socialization (Habitus) in Tension with Social Transformation (Project)

By and large, socialization theories of personality are considered discrete from transformational theories. Since most of what constitutes mainstream psychology is "social maintenance" in nature, most theories are conservative in nature; that is, they express an overriding interest in what I will be calling identity, the social self, order, habit (i.e., preconsciousness), ideology, and determinism. A psychology based on prediction and control would naturally move toward this side of the polarity. Transformational theories have been much rarer in psychology. A prediction and control mentality probability finds difficulty in dealing with change; therefore this lacuna ensues. The classical works on person-

TABLE 2.   Integrative Polarities of a Personal World

Socialization		Transformation
(Habitus)	←——→	(Project)
1. Social self (me)	←———→	Intentional self (I)
2. Identity	←———→	Individuation
3. Order (reproduction)	←———→	Change (transformation)
4. Preconscious	←———→	Conscious
5. Past (history)	←———→	Future
6. Ideological symbols	←———→	Utopian symbols
7. Determinism	←———→	Freedom
	Dialectical Integration	

al transformational theories have been theories of "conversion." The classic work in this genre would have to be William James' *Varieties of Religious Experience* (1902/1936). In this book, James attempts to come to terms with the reality of radical personality transformation through religious experience. The brilliance of this work has never been truly appreciated in American psychology, although the work was created by an American. Robert Lifton's (1974) studies on thought reform in extreme social situations seems to capture the spirit of this earlier work, but for the most part this genre has been more of interest to theologians than to professional psychologists. In elucidating the components of a personal world, I will attempt to hold the socialization and transformation aspects in dialectical tension. The focus throughout the exposition will be on the tension of what I will be calling "habitus and project." I am using the term *habitus* after Bourdieu and Passeron (1977), whose work complements my own. The notion of project is shared by several authors that I will shortly allude to. Table 3 poses in capsule form the questions posed in the tension of habitus and project.

Let me now explore the polar tensions of what I am calling "habitus" and "project" under the six polarities presented in Table 2.

1. *Social self (Me)—intentional self (I)*. One aspect of the development of a person in a personal world is the development of the social self (me) which Mead characterizes as the mirror image of the culture. Although the appropriation of the social self has frequently been characterized in a passive way (e.g. traditional social learning theory), I am departing from this tradition by emphasizing the agency of the self in the appropriation of a social self. In other words, the habitus of a culture is transmitted in a dynamic and reciprocal interchange between culture and person.

The intentional self is that aspect of selfhood which constitutes the possibility of human creativity (Winter, 1966). The social self, (the Median "me"), with its emphasis on reproduction, must be counterposed with an intentional self (the Median "I") if a legitimate understanding of human freedom is to be sustained as a viable concept in the interpretation of human action. It is clear

TABLE 3.   The Habitus–Project Distinction

Habitus	Project
How does culture reproduce itself?	How does culture transform or change itself?
Another way of putting the question is to ask: How does the personal world become integrated into a larger cultural world so as to reproduce (socialize) that world in the new generation?	Another way of putting the question is to ask: How is the personal world changed by new interactions that move away from the past so as to change the culture (transform) in some essential characteristics?

that human action must be seen in relation to real factors which condition its reproduction (i.e., continuity). Most of what constitutes "mainstream psychology" is geared to capture the habitus in a microcosm. The personal world is not only governed by the past but also projected into a future (Winter, 1966). The notion of "project" expresses the importance of the intentional futurity of human action. The notion of project is more than simply the goal-directedness of human action in the limited sense of purpose. In its broadest and richest sense, the project is the *constitutive* role of the self in action (Winter, 1966). We have outlined this constitutive role in rudimentary form in Chapter 2, in our discussion of human expression. In that sense, the project is characterized as a *conscious* expression, intentional in nature and embodying intentions, responsibility, and significance. Humans define themselves by their projects (Sartre, 1968). If this assumption is accepted, then it has important consequences for any interpretation of human actions. Laing and Cooper (1964) bring this to the forefront in their discussion of the notion of project:

> Any simple observation of the social field should make us realize that the relation to ends is a universal structure of human enterprises and that it is on the basis of this relation that men understand actions and institutions. Comprehension of the other is achieved only through a realization of the ends of his acts and projects. One may watch a man working from a distance and feel that one does not comprehend what he is doing, until one realizes the end at which he is aiming in his labour so that all his different movements become unified in the light of the end. Ends are not mysterious entities or some sort of appendages to acts. They simply represent the depassment of the given in an act which passes from the present to the future. (p. 64)

Human action, as a constitutive act, looks forward in its projects (Winter, 1966). Sartre (1968) maintains that a simple inspection of the social field leads one to discover that the relation to ends (i.e., projects) is a permanent structure of human enterprises. If this is true, then to ignore the projective character of human action is to ignore something essential in human activity. The intentional self or I, therefore, represents the voluntaristic and creative aspects of human action. Mead and Dewey spoke of the intentional self in organic and biological terms. They likened it to impulse. It is impulsive within this interpretation because the action of the intentional self (or I) was seen as transcending or going beyond social norms (the me). Freud would talk about this as the dialectic of the id with the superego. This formulation puts the creative aspects of human actions as an asocial endeavor. This is accomplished in an unreflective manner, since most social theory treats the social order as it is presently constituted as normative. Thus any human action which goes beyond a presently formulated consensus (social order) can be considered impulsive in this light. In order to legitimate the notion of project as constitutive of human action, it is necessary to contrast our discussion of order and reproduction with a conception that gives credence to the possibility of social transformation.

The dialectic between habitus and project is essential to human development. The order that culture conveys provides the basis for a sense of identity. The sense that this order is not an absolute provides the basis for a sense of individuation and uniqueness. These aspects must be seen as polar tensions rather than distinctive categories. Therefore, identity and individuation are two faces of human development. We turn now to that polarity.

2. *Identity—individuation.* Erikson (1959) defines "identity" as a sense of self-sameness in the process of change. The active and generative quality of one's history (i.e., habitus) allows a sense of "self-sameness" to be built up. If this is absent, an individual is said to be "identity diffused." Most psychological discussions of identity are formulated within the framework of individualism. In shifting the ground away from an individualistic framework, we will discuss identity in dialectical terms. Our discussion of identity will be in terms of class, gender, and ethnic identity. Because this is a departure from conventional psychological categories, it is necessary to elaborate. Let me give an example relating to class identity.

The habitus is a generative concept which attempts to account for the effects of collective history. Bourdieu (1977) elaborates on the class nature of habitus as follows:

> Since the history of the individual is never anything other than a certain specification of the collective history of his group or class, each individual system of dispositions may be seen as a structural variant of all the other group or class habitus, expressing the difference between trajectories and positions inside or outside of the class. (p. 86)

To talk about class identity, one must think in dialectical terms. Confrontation between agents of different classes is an interaction defined by the *objective structure* of the relation between the groups involved (Bourdieu, 1977)—for example, a superior giving orders to a subordinate or a professor discussing the relative merits of students in program planning. Confrontation between classes also brings out "systems of dispositions" within and between classes (Bourdieu, 1977). These systems of dispositions as generated by a class habitus can be seen in such areas as linguistic and cultural competence, which are embodied actively in a competence acquired by a particular history (Bourdieu, 1977).

Let me give a few examples of what these differential dispositions might be in a class society. First, let us look at a possible motivational dimension in class terms. The habitus here helps us to understand why a class society can remain somewhat stable through the history of its individual agents. Coles (1977), in a study of children who are born in families of privilege, detects in these children a pervasive feeling of "entitlement." Entitlement signifies that these children expect that things will happen for them and in their favor. Entitlement is a dispositional expression of those familiar with class, prerogatives, money, and power (Coles, 1977). An instance of this can be seen in an interview with a 6-year-old who knew that at 21 she would inherit half a million dollars. Looking at a working-class population of adults, one sees a different dispositional set that nevertheless solidifies and complements the entitlement of the privileged. Sennett and Cobb (1973), in an in-depth study of male working-class adults, indicate that they have a pervasive "feeling of not getting anywhere despite their efforts." There is also a feeling of vulnerability rather than entitlement in contrasting and comparing themselves to others at a higher social and economic level. This is not to say that working-class children do not *resist* the vulnerability discussed above (see Willis, 1977). The problem is that in many instances this resistance eventually is a rebellion that entraps them further in a rigid class identity. Lower-class children in a class-structured society come to see themselves as having "small futures"—the converse of the entitlement of the privileged. This dialectical dispositional movement between classes tends to solidify a class habitus.

The personal world must be seen within the development of a class-structured society. As Bernstein (1975) aptly points out:

> The class structure influences work and educational roles and brings families into a special relationship with each other and deeply penetrates the structure of life experiences within the family. The class structure has deeply marked the distribution of knowledge within society. It has given differential access to the sense that the world is permeable. It has sealed off communities from each other and has ranked these communities on a scale of invidious worth. (pp. 198–199)

Part of this scaling which is based on class identity is seen in the accumulation of "cultural capital" (Bourdieu, 1977). Here we are switching our focus from motivational to cognitive dispositions, which may be characteristic of different class identities. Bernstein (1975) gives one example of what I am trying to get at here in his study of "speech codes" between different social classes. The dominant speech mode of middle-class children is one that is characterized as an "elaborated code." The elaborated code is a speech pattern that reveals subjective intent of the speakers, shows a sensitivity to the implications of separateness and difference, and, finally, involves an orientation to a complex hierarchy for the organization of experience. By contrast, lower-class children reveal a code that Bernstein (1975) labels "restricted." In the restricted code, there is a speech form that discourages the speaker from verbally elaborating subjective intent; it progressively orients the user to descriptive rather than abstract concepts. It is necessary here to make clear the nature of these sociolinguistic patterns. Bernstein (1975) makes it clear that these patterns are independent of nonverbal intelligence measures. The issue here between these class patterns is not fundamentally that of intelligence. From the perspective that is being developed in this book, one should not conclude that these differences are organically related to class placement. Seeing these results within a personal metaphor, we would contend that these code differences are a function of different social-structure and class constraints. They reveal just as much about the class structure as they do about the members of these individual classes. We would contend that these codes could be changed if the class structure (i.e., class habitus) that buttresses these differences actively were changed. Under present class conditions, the codes are illustrative of one group naming the world (i.e., middle-class elaborated code) and another group having their world named for them (see Freire, 1974).

To talk about personal identity in class terms is to accent the relational quality of identity formation. Here I would like to stress that the notion of class identity should not be turned into an absolute. Ethnicity, race, sex, and so on must also be coordinated and integrated into a conception of social identity if this concept is to be productive. Class being equal, there are considerable differences between blacks and Puerto Ricans that demand a subtlety in any class analysis of character structure. Gender identity is another important dimension to be explored even when class identity is equivalent. In the past, however, class differences were mostly ignored in the study of these other differences (e.g., race). The identity of the self is therefore accomplished in historical situations. The consolidation of a personal identity is therefore carried out within real historical conditions that operate in class gender and ethnic structures. To interpret personal identity with these dimensions in mind will demand a considerable amount of theoretical and research sophistication.

The polar dimension to *identity* is the development of *individuation*. Individuation expresses the fact that the process of personal development expresses the possibility of some *uniqueness* and *resistance* to a dominant culture. If culture only reproduced itself in new members, then there would be no creativity in human institutions. This, of course, is not true. The *intentional self* is not only the receiver of cultural constraints; it is also a transformer of the institutions that mold it. A "new generation" always challenges its elders to be different from the past (i.e., habitus). When the balance is weighted toward cultural identity, then cultural conformity predominates. When the balance is weighted toward individuation (i.e., projection) then a condition of radical social change is in operation. The balance of identity and individuation is always delicate. Some epochs are characterized by one or the other of these polar aspects. The 1960s in North America, for example, represented an epoch weighted toward individuation. There was serious challenge to the status quo in class, race, and gender relations. The 1980s appear to have swung toward a more conservative orientation. If an orientation or movement that challenges the status quo is operating, then "individuation" becomes easier to interpret.

Individuation expresses the possibility of the transformation of cultural patterns. The notion of *project* indicates the real possibility for individuation that enhances the possibility of cultural transformation. The project means, in essence, that the past can be transcended. The intentional self is, therefore, an individuating aspect of the self. What we mean here is simply that the self is not only molded by culture (i.e., habitus) but also molds and transforms culture (i.e., project). In talking about individuation, I do not want this concept to be confused with the notion of individualism which I have already critiqued. Individuation, like its polar opposite, identity, is worked out and projected within a culture. It is expressed in the *resistance* and *transcendence* to the status quo. This individuation can be expressed differently across class lines. For example, Keniston's *Young Radicals* (1968) were middle-class students who rejected the normal entitlements of their middle-class background. In the opposite direction, Chicano farm workers made demands for an "entitlement" of the American pie, which was not part of their traditional pattern of interaction. What we are trying to indicate here is that these aspects must be seen dialectically and within historical conditions. The polarity of identity and individuation, in essence, is part of our larger developmental dialectic that I am calling the "polarity of order and change" (transformation).

3. *Order (reproduction)—change (transformation)*. The habitus is the dynamic process of becoming a "me." It is dynamic because it is a generative rather than a passive process. Looking at it in dialectical terms, it is the dialectics of the internalization of externality (the other) into the self or incorporation and the externalization of internality (self-projection or objectification; Bourdieu, 1977). The habitus is, therefore, the product of cultural appropriation and incorporation necessary for these products of collective history (i.e., culture), the objective structures (e.g., language, economy, etc.) succeeding, more or less completely, in reproducing themselves in the form of durable dispositions (i.e., social self or me; Bourdieu, 1977). This social self (me), as we are calling it, is both a product and a producer of a social environment. The social self or habitus must be seen in dynamic structural terms. The structural dialectic is developed by Bourdieu (1977) as follows:

> The structures constitutive of a particular type of environment (e.g., the material conditions of existence characteristic of a class condition) produce *habitus*, systems of durable, transposable dispositions, structured structures predisposed to function as structuring structures, that is, as principles of the generation and structuring of practices and representations which can be objectively "regulated" and "regular" without in any way being the product of obedience to rules, objectively adapted to their goals without presupposing a conscious aiming at ends or an express mastery of the operations necessary to attain them and, being all this, collectively orchestrated without being the product of the orchestrativy action of a conductor. (p. 72)

The structured ordering that is generative in nature has its eye on the past; one could call it the active production (i.e., reproduction) of cultural memory. The social science literature, such as it is, attempts to account for social order and maintenance (Touraine, 1974). In fact, the mainstream definition of psychology as "prediction and control of human behavior" is, of its very nature, a social maintenance (order) endeavor. Within this definition, change must be predictable and orderly in nature in order to have a science that espouses social determinism. The idea of an intentional self (I) sets clear limits to a social science based on social determinism (Winter, 1966). When the "intelligibility of human action" is interpreted in a social–deterministic framework, human creativity becomes anarchic and impulsive. It is considered as disorder rather than some attempt at social transformation. A critical interpretative framework must embrace personal and social change (i.e., transformation) as a legitimate aspect of human action. Sartre (1968) maintains that the most rudimentary action must be determined both in relation to real and present factors that condition it (i.e., order and reproduction) but also in relation to certain objects or purposes still to come that action is trying to bring into being. The idea of the surpassing of the past and present by human action, therefore, establishes transformation and change as a legitimate rather than disordered aspect of human action.

4. *Preconscious—conscious.* The habitus is unconscious as a generative structure. The structures of the habitus are therefore tacit in the sense of what Polyanyi calls the "tacit dimension." The unconscious structures mean the forgetting of history. Like tying one's shoelaces, one assumes from facility that we must have been born with that ability. In reality, by watching a child in his or her nascent movements at this task, we realize that shoe tying is a learned process that has become second nature. Bourdieu (1977) sees the habitus as a generative structure that is made by a history and, in turn, makes history.

> The unconscious is never anything other than the forgetting of history which history itself produces by incorporating the objective structures it produces in the second natures of habitus. . . . in each of us, in varying proportions, there is a part of yesterday's man; it is yesterday's man who inevitably predominates in us, since the present amounts to little compared with the long past in the cause of which we were formed and from which we result. (pp. 78–79)

Personal histories, then, will mirror some aspects of a larger cultural history, and history, instead of being called "dead weight," will be seen as dynamically generative, reproducing itself in predictable cultural ways (i.e., mores and customs). The habitus, then, can be said to be a mediating process generating social practices by individual agents which is culturally reproductive. This is done without explicit reason or signifying intent, making a particular social practice coming from a particular individual appear "sensible" and "reasonable" (Bourdieu, 1977).

The intentional aspect of the self is the conscious producer of human expressions. Consciousness implies the deliberative and reflective actions of an intentional agent. Intentions should therefore be understood as that dimension of human action which is consciously held in mind as orientation toward a goal. The project as a conscious act is, therefore, purposive in nature. Intentional or purposive acts are those which the agent knows or believes can be expected to manifest a particular quality or outcome and in which this knowledge is utilized by the actor to produce the quality or outcome (Giddens, 1977). We have discussed this aspect of the project in Chapter 2 at some length. What I would like to stress here is the dialectic of unconscious and conscious aspects of human agency. Conscious intentions are built on the sediment of past intentions, which are now preconscious. Habit allows for a greater economy in human intentions. Habit allows one to free oneself from constant deliberation in order to project new intentions. In projecting new intentions, old habits may be confirmed or rejected. The building upon habits (preconscious) and the transforming of habits (projects) expresses the dialectic of the conscious and unconscious. It is the *present* embedded in the dialectic of the past and future.

5. *Past (history)—future.* In one sense, the habitus is the rehearsal of what has been in order to keep it happening. It is, therefore, the *past* being reproduced in the *present.* It is history turned into nature (Bourdieu, 1977). Culture is, by definition, second nature. What is natural for humans is that they are, by definition, second natures (i.e., cultural animals). In no other species is the rehearsal of social history so necessary for basic survival. Reproduction of culture is part of the makeup of humans.

The human being is not only a past or a present (a history made) but is a history in the making through his or her projects. Merleau-Ponty (1963) puts this in the following way:

> What defines man is not the capacity to create a second nature—economic, social or cultural—beyond biological nature, it is rather the capacity to go beyond created structures in order to create others. (p. 175)

The intentional futurity of human action should not be seen in a futuristic vacuum. Humans create their futures in history and create it while being weighed down by a past which is, in some very important way, determining. The reproduction of the past in human action serves the maintenance of a specific social order. The projection of human action into the future allows one to consider the possibility for the transformation of social orders.

6. *Ideological symbols—utopian symbols.* Mannheim (1953b) gave a specific definition to the concept of "ideology" by noting the discovery that emerged from political conflict, namely, that ruling groups can, in their thinking, become intensively involved in reproducing (i.e., maintaining) a particular social order so as to become totally unconscious to certain facts which would undermine their sense of domination. There is an insight implicit in the concept of ideology which alerts one to the collective distortions of habitus; that is, in certain situations the collective unconscious of certain groups obscures the real condition of society both to itself and to others and thereby stabilizes it (Mannheim, 1953b). The work of Marx is an extensive excursus on how class relations and domination produce a collective ideology which is distorted by the unconscious nature of class domination. The habitus is not simply the "collective unconscious of history"; it is the reflection of the power relations productive of a

particular social system. One can, therefore, speak of a "habitus of capitalism," of which we shall say more later. The habitus is, therefore, a generative unconscious production of symbols and linguistic systems that extol the nature of the present social order, actively seeking to maintain it. Ideological symbols are, therefore, symbols of loyalty to the present order, extrolling the worth of that order. Maintenance of these symbols is first projected in the family unit, but in complex societies they are solidified by more impersonal institutions such as the school, mass media, and the state. Ideological symbols (e.g., pledge of allegiance to the flag, etc.) extol the past and help in the production of social maintenance by indicating that the "present order" is the best of all possible worlds. These symbols are the "cultural capital" of ruling classes, where ideological symbols tend to buttress their domination (Bourdieu, 1977). The project, by contrast, moves toward the future and has the potential to challenge the total legitimacy of the past. The first indication of an alternative ordering can be seen in the child's use of negation. Two-year-old children, as a result of their great "no," are felt to be going through the "terrible twos" by parents and psychologists alike. What is implicit in this negation is a challenge to the powers that be (i.e., habitus). Resistance is the first indication that things could or should be otherwise than they are. Negation is a core element in the development of utopian symbols. Utopian symbols are symbolic images of the future that challenge the ultimacy and total legitimacy of the present order. Utopian imagination has a sensitivity to the breaking points of the present system and nourishes a longing for a new kind of social ordering (transformation); thus it exercises a significant role in social change (Baum, 1975). The absence of utopian symbols allows the present order to go unchallenged and results in a static state of affairs in which people become molded to what appear to be the fixed laws of the social system. The project, if it is to be authentic, incorporates utopian symbols that concretize alternative futures:

> Concrete utopians are images of the future that are grounded in athentic intuitions of the ills and contradictions of the present society. Concrete utopias negate the most oppressive elements of this society and present a vision of human life that, even if as such unrealizable, summon forth new ways of thinking and acting that could lead to social change. (Baum, 1975, p. 171)

The notion of project therefore accents the protensive movement of human action set in terms of a posited or fantasied state of affairs in the future perfect tense (Winter, 1968).

7. *The dialectic of determinism—freedom.* In commonsense terms, determinism says "what has happened before will happen again." Determinism is, in essence, the past ordering the present. The habitus is a generative structure that allows some predictability in the social order as well as one's individual personal life. If you know a person or a group history, you will have some orderly sense of

what may happen in new situations. We are not using *determinism* here in terms of mechanical human causality. As we will see when we discuss the notion of project, humans are formed by history (i.e., determinism), but they also alter the course of their histories (i.e., freedom). Humans, as cultural animals, are determined (i.e., made by history) and determining (i.e., make new history). Determinism produced by accumulated history (habitus) allows for the possibility of a sense of identity. Change opens the door to individuation and therefore freedom.

If the commonsense notion of determinism is "what has happened before will happen again," the commonsense version of freedom is that "what has happened before need not happen again." It is not necessary to hold a conception of freedom by throwing away a conception of determinism. A conception of freedom can mean simply that the past does not have absolute power over the present and future. Freedom means a certain openness to the future, where alternatives to the past may operate and possibly be brought to fruition. Sartre (1968) is more lucid in relating a structure of freedom with the notion of project:

> Man defines himself by his project. This material being perpetually goes beyond the condition which is made for him; he reveals and determines his situation by transcending it in order to objectify himself—by work, action, or gesture. The project must not be confused with the will, which is an abstract entity, although the project can assume a voluntary form under certain circumstances. This immediate relation with the Other than oneself, beyond the given and constituted elements, this perpetual production of oneself by work and "praxis," is our peculiar structure. It is neither a will nor a need nor a passion, but our needs—like our passions or like the most abstract of our thoughts—participate in this structure. They are always "outside of themselves toward. . . ." This is what we call existence, and by this we do not mean a stable substance which rests in itself, but rather a perpetual disequilibrium, a wrenching away from itself with all its body. As this impulse toward objectification assumes various forms according to the individual, as it projects us across a field of possibilities, some of which we realize to the exclusion of others, we call it also choice or freedom. . . . (pp. 150–151)

The notion of project or any semblance to it has been absent from the vast corpus of psychological research. This is partly due to the fact that psychology as an institution has been mostly involved in the maintenance of the social order (i.e., socialization). In declaring itself a deterministic science, it eclipsed the notion of freedom. In venturing a conception of freedom as an essential component of critical interpretation, we do not exclude determinisms. Rather, it is necessary to incorporate freedom within a dialectic of freedom and determinism. I would contend that there is no possibility of freedom without a solidified past (i.e., habitus). At the same time, there is an absence of freedom when the past is made an absolute. The transformation of the past (habitus) into possibilities for alternative futures (i.e., projects) is the very essence of human freedom. Humans are certainly made by their past, but, by their projects, they are not condemned to it.

The Personal World as a Located Cultural Form

We have just treated the personal world as a set of dialectical polarities that hold in tension the concepts of socialization and social transformation. Under the tension I have discussed two core notions, the habitus and project, and outlined in dialectical fashion seven polar dimensions that characterize these notions. I would contend that these polarities are the dynamics of culture. The metaphor of the personal assumes a concept of culture within its formulation. The personal world is a cultural world. This is why interpretation is important, since the cultural world creates symbols that need deciphering. A conception of the human as a cultural animal cannot be formulated within a paradigm of mechanism or organism. Shotter (1975) maintains that:

> Man is not just man in nature, he is man in a culture in nature. And a man's culture is not to be characterized in terms of objective properties like all the other things that he sees in nature from within his culture. The culture from within which a man views the world and deliberates upon how to act in it structures his consciousness, and can only be characterized in terms of his beliefs. . . . Thus, if there is a key to the contemporary phase in the development of the human sciences, it is this: there is a third term to the relationship between man and nature, *culture*, which is not genetically inherited but communicated to man after birth as a "second nature." It is this third term that psychology, in its attempts to be scientific, has ignored. (p. 136)

I would like to extend the notion of the personal world to that of a located cultural form. In doing this, we move away from any facile idealization of the personal metaphor. The contention is that the personal world and all its relationships are embedded in a specific history with all of the idiosyncrasies and limitations entailed therein. The personal world is, therefore, a real historical event rather than some abstract formulation of an ideal person. Therefore, when we speak of a personal world, we are essentially talking about an identifiable located cultural form (see Willis, 1977). The essence of that form is constitutive relationships—that is, the particular way a social group is connected to the objects, artifacts, institutions, and systematic practices of others that surround it (Willis, 1977). The individual of a personal world exists as much through similarity as through difference. The paradox here is that individuation is partly achieved by becoming similar to persons in one's group and different from those outside one's group. These constitutive relationships need not imply conscious purpose or intention on the part of the social group (Willis, 1977). In other words, people may share a common world (cultural forms) but do so habitually, unconsciously, and unintentionally. At the same time, parts of that world may be coconstituted intentionally by members. In other words, we are saying that a personal world is a located cultural form created out of habit and intention (i.e., the dialectic of habitus–project). The dialectics of habit and intention allow a personal world the possibility of identity and change at the same time. To talk of the personal world as a located cultural form is, therefore, to talk of a dynamic totality. The personal world as a cultural form implies real exchanges and material dialectics between the structure of human subjectivity, its characterization sensibilities, and its context (Willis, 1977). It is, therefore, a set of relations embedded in material–historical realities. Thus, the personal world is not an abstract idealistic concept; rather, it is the world of groups such as Poles and Irish, blacks, workers, bankers, women, and so on.

What constitutes the boundaries of a personal world? For our purposes, a personal world is an identifiable group of individuals whose relations with one another can be characterized as a reciprocity of equals (relatively speaking) and who center their relations within themselves both materially and symbolically. It is important to distinguish the personal world from larger structural totalities in which it is enmeshed. For example, the commercial system, productive relations, ideological and political formations constitute these larger structural totalities which can and are studied as independent events by other disciplines (e.g., political science, history, economics, sociology, etc.). It is understood in our treatment that these larger structural totalities are not independent of a personal world and cannot be studied as if they were totally discrete facts. This is the topic of the next chapter. In relationship to a personal world, these larger structural totalities can be understood to intervene, redirect, or block the possibility of intent, pure expression, or unity of the personal world. The personal world must be studied dialectically in relation to larger structural totalities. The role of a psychological interpretation of the personal world is simply to accept the microsocial world of the personal without making an individualistic fetish of it as does traditional psychological theorizing. We do not talk about persons in the abstract. Rather, we talk about black persons who are black Americans, French people who are French Canadians. Persons are persons only within a cultural world. There is no person if there is no cultural form. I am, therefore, talking about a concept of the person which is historical in nature. Considering biography as history, I would propose that a critical interpretation of the personal world is an excursion into biography. A psychology of this vintage enables us to focus on the life span (i.e., biography) of the individual as a site of psychological explanation (Hudson, 1972). To do biography is to do an interpretation of a personal world. Critical interpretation is the ability to identify dialectically those factors in a personal world that accent the development of human freedom or detract from that endeavor (i.e., domination). Critical biographical work involves locating biographical material within larger structural totalities. We will introduce these totalities in the next chapter in discussing the notions of class, gender, and ethnicity.

SUMMARY

We have now made the turn out of egocentricity and into culture. This was achieved in a series of steps. Our first task was to distance ourselves from the impediment of individualism, which leads into egocentrism at the level of theory and practice. The next step was to indicate that *intentional expression* at an egocentric level of analysis (see Chapter 2) is the *project* in the personal world. This notion of project was one of the two major components of the personal world in our introduction of the notions of habitus–project. From our treatment of the dialectics of habitus–project in the personal world, we were able to make the transition to the notion of the cultural form. We can now say that the personal world is a cultural form. This equation prevents us from slipping back into egocentric analysis. Willis (1977) makes my point explicit:

> The group is the smallest unit of cultural existence. It is the smallest unit of resistance against the dominant culture, and the immediate form of protection of, and influence on, *particular* subjectivities. It is the pivotal site for the development of wider social and cultural practices. Though the *material* of experience is necessarily often collected (as a research act) from the individual, and though the expression and form of this experience may be personal, the real generator of this experience is the interaction of the group within a specific social and material matrix of symbols, expectations and conditions. Though subjective, then, the focus of my interest is on the *subjectivity* of symbolic systems, actions and values, *not* on the individual person; or more correctly it is only on the individual as he lives out personalized versions of collective strategies. (Willis, 1977, p. 178)

I have given this long quote because of the clarity of focus it gives for the direction I am taking in this book. It is now time to explicate in much greater detail the dialectical relationship between the personal world as cultural form and the larger structural totalities class, race, and gender.

4

PERSONAL WORLD, CULTURAL FORMS, AND THE STRUCTURES OF CLASS, GENDER, ETHNICITY, AND AGE

INTRODUCTION

We have just completed a linkage between the personal world of human action and that of the cultural form in that we have built a bridge between the personal world of human intentions and cultural action, an expressive action governed by a dialectical tension identified as habitus–project. In this chapter, I will attempt to sensitize the reader to structural relationships of power and their impact on the personal world. By structural relationships of power, I mean power between agents rather than a specific set of powers supposedly located within agents. These relationships of power will link the personal world of cultural forms to the structures of class, gender, ethnicity, and age.

POWER, AGENCY, AND DOMINATION

In English, the word *power* is derived from the Latin word *potere*, of which the root meaning is "to be able to." As a verb, it cannot stand alone and needs completion. Thus, one does not say "I am able to" but "I am able to jump 3 feet high." Strictly speaking, at the level of personal agency one could say that power is a condition where one is "enabled." I would contend that this is a condition of personal agency. From the point of view of individual agency, for the moment one can say that personal powers are present when a human act is identified. In Chapter 2, I identified a human act as having the characteristics of consciousness, intentionality, intentions, responsibility, and significance. What was clear from my treatment in Chapter 2 was that personal agency, in my definition, did not locate agency within agents but rather between agents. This was the reason for my rejection of mechanical and biological metaphors. Therefore, under the personal metaphor I would locate power as a condition between agents. I have maintained that the personal world is a world of mutual reciprocity and equality between agents (I–thou). Thus, there exists a mutual enhancement of personal powers when the personal world and its agents are operating cointentionally. By this, I mean that the world of an agent (I) has its conscious, intentional goals (intentions) that are responsibly significant, enhanced by the other (thou) and vice versa. This type of cointentionality is called cooperation. So mutual agency implies cooperation. When a condition of cooperation is vitiated or absent, one calls this an impersonal relationship. This is a condition where one agent (or agents) treats another agent (or agents) as an object (I–it relationship). We call this relationship inhuman or oppressive. It is a structural condition that will be referred to as *domination*. A condition of domination is operating when the agency of one person or group eclipses or masks the agency of another person or group. Thus, the power of one is at the expense of the other. If the personal relationship could be referred to as *cooperative*, the appellation of *oppression* characterizes a condition of domination. Oppression operates in the following manner: the agency of one person or group (i.e., consciousness, intentionality, intentions, responsibility, and significance) denies the possibility of the other person's or group's consciousness, intentionality, intentions, responsibility, and significance. This situation for those dominated turns agent conditions listed above into their opposite. Thus, for consciousness we have repression, for intentionality we have withdrawal, for intentions we have an absence of goal orientation or rootlessness, for responsibility we have dependency, and for significance we have meaninglessness or anomie. A relationship of domination creates *powerlessness* in the dominated person or group. The most extreme form of this condition is depersonalization, where one person or group treats another person or group as a total object (I–it). This asymmetry in power was described by Hegel as the "master–slave relationship."

Structures of Domination

From my own experience, it is clear that some groups dominate others: men dominate women, adults dominate children, and capital dominates labor. This is not a "fact" *per se*, but I believe that what I am saying is reasonable to assume and can be demonstrated by argument. For example, Wilden (1980) would say these relationships of domination are "real." In other words, they have stability and "social integrity." They are conditions of inequality in social relationships. Wilden (1980) conjectures that there are two ways to camouflage these relationships of inequality. The first one is "imaginary symmetrization." To symmetrize is to venture that capital is equal to labor, women to men, blacks to white, children to adults. Thus, differences between these oppositions are considered accidental and temporary. The apparent unequal relationships can be altered through education, incentives, and so on. I call this liberal ideology (see Chapter 3). The second camouflage is "imaginary inversion" (Wilden, 1980). This inversion is a paranoid set that indulges in stereotypes and agitation. Thus, labor has gotten too big, women are now dominating men, and so on (Wilden, 1980). This is the camouflage of fascist ideology. Both liberal camouflage or ideology and the fascist deny the relationships of inequality to which we have originally alluded. By and large, conventional social science and, for our purposes, conventional psychology indulge in some version of the liberal ideological camouflage (see Sullivan, 1977a, 1977b, 1980a). Specifically, psychologists take structural relationships of power such as capital over labor, men over women, and change them into intrapsychic phenomena. Thus, women's inequality in relation to men can be seen as precipitated by a motive called "fear of success" (Horner, 1970). Black children's inequality to whites can be seen as cultural deprivation. As we shall see later, a critical–interpretative psychology will attempt to deal with these structural inequalities without liberal or fascist camouflage. What we must attempt at the outset is to indicate how these structural inequalities relate to one another and to the specific cultural forms that constitute the fabric and underpinning of any social order.

Domination in the Context of Cultural Forms

In Chapter 3, I indicated how the cultural world is embedded in a dialectic of order and change called the habitus–project tension. It is now necessary to link these concepts to the structures of domination and the dimension of power. Within the dynamic of habitus–project, domination can operate in two distinct but by no means separate ways. The first way is through the destruction and destabilizing of an ordered cultural form (i.e., habitus). Habitus or stabilized social relationships are the firm footings for cultural projects. The second way is through the direct repression of the projective character of a specific cultural form (i.e., project). Repression within this context destroys the dynamic forward movement of a cultural project (i.e., creative project). Domination is achieved by operating simultaneously on both stabilizing and dynamic core of specific cultural forms. Its end product is disabling (i.e., powerlessness).

CLASS AS A STRUCTURE OF DOMINATION

I now enter an area that conventional psychology ignores but that is essential to my treatment of the personal world of specific cultural forms. This is "class analysis" in the tradition of Marxism. I have already alluded to a dominance relationship of capital over labor. This is the category of dominance in which traditional Marxism discusses class analysis. The notion of class, as a result of the profound thinking of Karl Marx, has been identified in its modern coinage with economic categories. It is reasonable to assume that economics would be pre-emptive in a trade society like our own where class relations have a clearly identifiable economic dimension (Touraine, 1979). The complexity of class relations in modern societies makes it difficult to recognize clearly identifiable and recognizable social classes (Touraine, 1979). This does not detract from the importance of considering "class relations" as an important tool for social analy-sis, not only for economic categories but also for social and political categories as well (see Bourdieu & Passeron, 1977; Edwards, 1979; Piore, 1973; Touraine, 1979; Wright, 1976). Touraine (1979) reasons that social class is not defined simply by its relations of production, for this class also has the capacity to manage the state, to improve its ideology, and to repress or alienate its adversaries.

To make "class" a preemptive category for the analysis of dominance rela-tionship is questionable and suspect. Feminists are quick to point out that there are two evils that predate capitalism, namely, racism and sexism, in a criticism of male-dominated socialist revolution (Morgan, 1970).

It is clear that sexism is a more universal category of dominance than class, since it is present in the gender relationships of both dominant and subordinate classes. Racism is also not restricted to the dominant class and is clearly visible in virulent forms in subordinate groups. Frequently racism is a characteristic of relationships between cultural groups in the subordinate classes (e.g. blacks and Puerto Ricans). Just how these relationships of dominance are related is not clear, and no social theory has been successful thus far in explicating their interaction. This makes a case for caution for anyone, like myself, who ventures into this complex area. Thus, a viable class analysis demands a considerable amount of sublety (see Braverman, 1974; Piore, 1973; Wright, 1976). It would be pretentious to attempt to develop the complexity of contemporary class analy-sis in brief. What is being attempted here is to embed our conception of a personal world and its cultural forms in class structural terms. There is no simple class called laborers, for example; rather, there are Chicanos, blacks, Italians, Irish, and so on. It is not important or useful to pinpoint a "class" in contempo-rary class analysis. What is important is to emphasize the hierarchical nature of social relations, which is part and parcel of our contemporary productive and consumptive apparatus. It is here that I wish to locate a critical interpretation of the personal world and its cultural forms. A critical interpretation of that world will be sensitive to the relations of power and domination that prevent and jeopardize the development of human projects for some groups of people be-cause of the projects of others. In short, we are talking of relationships of exploitation. This constitutes a radical departure from most traditional person-ality theory, which has heretofore existed behind the imperatives of class and power, curiously independent and separate from the economic forces operating in society. In a society like our own, one can venture that the economic order of monopoly capitalism is stabilized through the dominance of capital over labor. This is done in numerous ways; the two most prominent are the control that the dominant group has over the means of production and also through incentives that come from consumption of products. Consider the possibility of two classes or groups interacting with one another. Rather than be prematurely specific, let us say one class or group is dominant and the other class is nondominant. The dominant group is characterized as having a clear ideological solidifcation (i.e., habitus) and a consciously operating project. The dominant class has control over the ideology of production, the cultural realm, political and state apparatus, as well as educational institutions (see Kellner, 1978). The dominant class has the capacity to manage the state, to impose its ideologies, and to repress and alienate its adversaries (Touraine, 1979). The nondominant class has a diffused ideology (habitus) or one that conforms in essential features to the habitus of the dominant ideological group. The difference between these groups as regards class habitus is that the dominant ideology is perceived as solidified (a condition of solidarity) in the dominant groups and diffuse (nonsolidarity and division) in the nondominant group. The project of the dominant class is perceived as the cultural project. The dominant group is in hegemonic control of the cultural project. The nondominant groups experience the lack of an authentic cultural project. This is a condition of alienation or oppression for this class, because of the overriding character of the dominant class's project. The first inclination of a project of the nondominant class is the presence of a resistance to the project of the dominant class. A second solidification of resistance is attempting a solidarity and class habitus that is independent of the habitus of the dominant group.

Let me take specific instances now to solidify the position I am attempting to develop. In doing this, I am assuming that class relations in a capitalist society can, after Wright (1976), be analyzed in terms of three processes underlying the social relations of production: (1) control of labor power, (2) control of the physical means of production, and (3) control of investment and resources. Wright (1976) contends that the central class forces in a capitalist society—that is, the bourgeoisie and the proletariat—represent polar class positions within each of these processes. Moreover, the polarities are maintained and legitimated in the ideological sectors such as family, school, and churches (Althusser, 1971; Bourdieu & Passeron, 1977).

Dominant Class

Members of the dominant class experience life as agents. They have a "solidified habitus," which is congruent with the prevalent cultural consensus. Because of the prevalence of sexism across all classes, males are more likely to perceive of themselves as agents than are females. In dominant classes, women solidify male dominance by supporting male agency without participating in it.

The dominant class controls the productive sector and, in essence, the cultural project of capitalism is defined within this class. Capitalism shrinks the scope of the human project by restricting utopian images to profit and consumer images and also lessening the possible range of free acts. Specifically, this can be seen by an examination of some of the empirical work on the phenomenology of the class structure. Maccoby (1976), in a book called *The Gamesmen*, did an intensive study of the upper echelons of the executive world. The "gamesmen," as he calls them, have to compete. They thrive on competition and the solving of problems. They accept, in my view, the dominant ideology of capitalism and enhance themselves on the upward mobility that is there for those who compete and succeed. They have a sense of a project and feel that they are in control of things. At this level of corporate life, the gamesman is a man rather than a woman, which is indicative of the implicit sexism in capitalism. The women of the gamesmen stand behind their men, supporting their projects by raising the children and by moving themselves and their families wherever the "games" have to be played. The gamesman supports "meritocracy" and sees his own success as based on his ability to compete and win. This project is the wider cultural project of capitalism, with competition at the apex.

Does the gamesman learn his project, as it were, "on the job"? By and large, he is probably raised in a "culture of affluence"; the "cultural form" of the gamesman's personal world has some unique characteristics that begin to be nurtured in childhood. It is here that the family and school develop a character which will meld with the expectations of future occupational status (Bourdieau & Passeron, 1977; Connell, 1977). Connell (1977), interviewing children from upper-status suburbs in Australia, gives a taste of this forming of character by early expectations for university education:

> Boy 12: Well, I'd like to—my mother wants me to go to university, and I want to go myself, and I want to say be an engineer kind of—my mother thinks it too hard, but my mother wants me to be a doctor, and I'm not keen on that. I'd like to do something more exciting—my mother said an engineer gets one of the most salaries in the world. (p. 153)

And even in the next instance, where a girl from a dominant class thinks she is not clever enough, a confidence and family support is still perceived:

> Girl 15: I'd like to be a teacher, I think they do a lot better than some people do. I'd like to teach either primary or infants. I'm not sure at the moment, I'd only have to go to teachers' college. Dad wants me to go to university and get a Bachelor of Arts degree but I don't think I'm clever enough for that. . . . I'm going in for a test on Firday to the Guidance Department for an interview about occupations and you have to name the three that you'd like to do. (Which ones did you?) I put teaching first, and then if I couldn't do that I wouldn't mind being a child welfare officer or a social worker, cause I like children. . . . (p. 153)

Compare this statement with that of a lower-class girl of 13:

> Girl 13: Well, myself I'd like to be a kindergarten teacher—I've got four brothers and sisters younger than me and I get along with them real good, play games with them on weekends and I get along with children. I'm very patient with them, not like my sister she goes off her rocker (laughs) bashes into them or screams or something. That's what I'd like to be or if I couldn't do that, which I don't think I've got the brains really, the intelligence, you know you've got to go all through school and the university and college and such, then I think I'd like to work either in some sort of shop, or then I'd go in for a factory. . . . (Connell, 1977, p. 152)

It is clear from Connell's examples and others (e.g., Coles, 1977; Willis, 1977) that the class-structure expectations reveal themselves in cultural forms in childhood expectations of students and parents. Let us now look at this outcome in more specific detail in the nondominant class.

Nondominant Class

The experience of a sense of agency is a difficult endeavor for members of nondominant classes. This is not to say that a sense of agency is totally lacking, but, by and large, it is experienced in what might be called the private sector of the family. Here, it is usually the male head of the family who experiences some semblance of power. Outside of the family, in the larger structural realities of work and politics, the member of the nondominant class experiences social reality as arbitrary or clearly within the control of the dominant class, however that is perceived. Members of the nondominant class generally sense the feeling of being a pawn rather than an "origin" of their social reality (De Charms, 1976). This position is operant when the nondominant class concedes the viability and legitimation of the social power of the dominant class. Nascent resistance to the dominant social reality of the dominant class is the kernel of agency that, if organized and consolidated, can challenge the power of the dominant class. Resistance, if politicized, can be the beginning of a project that is emergent and transformative. Resistance in itself is precarious and—when not challenging the structural totality of the dominant class—is short-lived and destructive of solidarity (Willis, 1977). It is not only agency that is lacking in the nondominant class; the "ordered stability" (i.e., habitus) which is integrated into the lives of members of the dominant class is experienced as vicarious and arbitrary by the nondominant class, since there is the absence of true participation in the structure of the dominant social reality. Thus, when a large corporation develops a community, the community can thrive on the economic windfall, but that reality can be decimated by a shift of that corporation out of the area. The absence of decisive control over the corporation in the community's life, a community project, leaves the ordered stability necessary for community life always precarious. It is here that corporate economic decisions have profound psychological consequences (Brenner, 1976). Let me now quote some empirical studies that attempt to elucidate the attempts at integration of the habitus and project of nondominant class members.

Sennett and Cobb (1973), in a book called *The Hidden Injuries of Class*, attempt to construct a quasiphenomenological account of adult working-class males. "Hidden injuries" represent the "personal cost" of the class structure to the psychological lives of these working-class males. What these authors try to demonstrate is how the "victims of a class structure" come to blame themselves for their plight. They are men who are not happy in their jobs, fear upward mobility because of peer rejection, and whose only solace is the fact that their children may be spared the working-class life and the boring jobs that they are required to do. Their major form of transcendence is the family, where consumption of commodities becomes the trade-off for alienating work (Sullivan, 1980a). Sennett and Cobb (1973) demonstrate the total hegemony of the class structure and its power to mute any sense of project in these workers, who blame the absence of project on their own previous educational histories or lack thereof.

There is a depressing tone to this study, and it stems from the apparent total absence of any resistance of these workers which might challenge the nature of their social realities (i.e., dull work) in terms other than that of their own shortcomings. One must question these authors' interpretative horizons, which seem to totally ignore the possible resistance of the working-class males they study. It would be important for a critically interpretative psychology not to miss signs of resistance to class domination, which are potentially transformative in nature. Aronowitz (1973), in a study of automobile workers, would appear to challenge the total passivity of working-class men in highly specialized and boring jobs. The automobile workers in the town plant of General Motors indicate a set of social relationships on the shop floor that are filled with rebellion and acts of resistance. It would be prematurely optimistic to name this kind of resistance a transformative project, but one should not ignore its nascent possibilities. The line workers resist by struggling to organize the production process in their day-to-day life. They do this at various levels of production by attempting to modify and control timing, pacing, and demands made upon them. They also appear to develop a certain camaraderie in their social relationships, creating rituals to break the monotony of their work routines. In general, this is accomplished through humor, small acts of resistance, and acts of outright sabotage that challenge the predefined authority of the workplace (Willis, 1977).

Willis (1977) elucidates very clearly the precarious nature of lower-class adolescent resistance in his ethnorgraphic analysis of social reproduction. His methodological technique maps out the reproductive path where working-class boys come to accept the seeming inevitability of working-class jobs. This, interestingly enough, is partially accomplished by the rampant sexism prevalent among these English working-class youths. In working-class youth, the value of manual labor is tied up with these youths' sense of masculinity (Willis, 1977). One interesting aspect of Willis's sample is the boys' active rebellion against the structure of the school system, in its institutional assault on their mode of dress, their language, their life-style, their sensuality. An active resistance to this assault is apparent in these students' rejection and ridiculing of the messages and demands of the school's overt and hidden curricula. Willis (1977) is very cognizant of the nonpolitical nature of this resistance, which is not directed to the structural dimensions of class, and ultimately determines that this type of resistance leads to entrapment rather than the transformation of institutional life.

What is important for an interpretative psychology of critical import is to sense the effect of the class structure on the personal worlds of cultural forms located within that structure. As with the Aronowitz (1973) and Willis (1977) observations, it is necessary to be sensitized to nascent forms of projects in dominated cultures. A critically sensitive psychology will not simply interpret this resistance as deviance, as is the case of most mainstream psychological interpretations of lower-class cultures. At the same time, this sensitivity must countenance the ambiguous nature of resistance in social positions which are lacking in social power, as are members of nondominant classes.

This leads me now to a summary statement on the nature of class structures and its dialectical relationship with the varied cultural forms that constitute to the dynamics of the class structure. We have established that a class structure is based on economic disparities which establish the dominance of one class over another. This domination is not simply economic in nature. It is maintained and legitimated by family, church, educational institutions, and state structures. It is also accomplished through the structures of ethnicity, gender, and age, albeit in a complex dialectical manner. All of these institutions combine to form the basis of a cultural totality, which we have labeled the "hegemony" of a capitalist class structure. The most pervasive and essential (but by no means exclusive) motive in a capitalist class structure is that of profit and capital accumulation. Profits are systematically accumulated more in the dominating class, and the legitimacy of this enterprise is buffered by the ideological structures of education and the state. There is a seeming naturalness and inevitability in this complex process, which usually goes unquestioned (Gramsci, 1971; Williams, 1976a). It is in this situated world of economics and ideological legitimation that the personal world is lived as a cultural form. Our contention in this chapter is that the cultural forms of the personal world are systemically influenced by the dynamics of the class structure. In schematic and summary form, let us look at this class dynamic when the class structure is stabilized and legitimated and when the class structure is being substantially challenged by the nondominant culture sustaining oppositional forms. The former dynamics will be called "formative" and the latter will be called "transformative." In the transformation phase, the oppositional structures of the nondominant class threaten the hegemonic structures of the dominant class. This can either be a condition of what is popularly called "reform" or "revolution." Tables 4 and 5 summarize the dynamics.

Having now set out a summary in tabular form of the dynamics of the class

TABLE 4. Hegemonic Stability (Formative Dynamics)

| Cultural Form of the Dominant Class | Cultural Form of the Nondominant Class |
Habitus	Habitus
(a) *Social self*—a solidarity is experienced with one's class. This solidarity is buffered by a clear sense of identity . . .	(a) *Social self*—although there is a "solidarity present because of common fate," it is not buffered by a strong sense of identity . . .
(b) *Identity*—in which one sees a close set of ties with one's parents and peers, which is integrated into an order . . .	(b) *Identity*—is diffused because the cultural dynamics of class and because the inadequacy of social power mutes a strong identification with one's parents and peers (i.e., father wants son to be different from him).
(c) *Order*—which is backed up by the legitimacy engendered for this class in the family ties, educational system, and state apparatuses. The cumulative residual of this ideological order is preconscious . . .	(c) *Order*—the experience of the social order appears arbitrary, and this class form is not legitimated by family ties, the educational system, or the state apparatuses. The school system slots members of the class for less prestigious educational outcomes, and this is done in such a fashion to allow members to be preconscious . . .
(d) *Preconscious*—so as to appear natural and inevitable rather than appropriated or consolidated. This preconscious consolidation is organized in a clearly ideological . . .	(d) *Preconscious*—of the systematic nature of this legitimated exclusion, thus appearing natural or inevitable and probably the best thing for all involved (i.e., complicity in their own oppression).
(e) *Ideological*—manner which justifies this seeming inevitability as eternal and historical and possibly biologically generated. Thus a determinism . . .	(e) *Ideological*—systems operate in this class by the acceptance of their exclusion from power and prestige as external and ahistorical (we always have been this way—we will always be the way we are now).
(f) *Determinism*—is experienced as to the natural location of one's cultural form in the class structure. This is the experience of natural entitlement, which allows no challenge to the present cultural hegemony.	(f) *Determinism*—is experienced as to the natural location of one's cultural form in the class structure. This is the experience of natural disenfranchisement which presents little challenge to the present cultural hegemony.
INTEGRATED	ALIENATED

Project →

(a) *Intentional self*—a sense of agency and congruence with the larger cultural project is prevalent. Children and adults in dominant social positions perceive themselves as originating their situation or fate. This allows for a sense of individuation . . .

(b) *Individuation*—which is, in essence, the sense that there is a place in the world for me which I, through my agency, bring about and originate. Change . . .

(c) *Change*—in a dominant position is experienced as orderly and predictable. The powerful sense of agency stemming from a congruence with the larger cultural hegemony is experienced as possible but systematically in the hands of the agent.

(d) *Conscious*—actions are governed by the directives of the larger cultural project. Under monopoly capitalism, these conscious decisions are governed in a superordinate sense (but not solely) by the profit motive.

(e) *Utopian*—images of change are always governed by reform rather than revolutionary visions. The utopian image in the dominant class is gradualistic and progressive and opposed to revolutionary and system-destabilizing freedom . . .

(f) *Freedom*—is experienced as the sence of agency, which comes with accomplishing the tasks of the larger cultural projects. It is also the absence of impediments to the agents' sense of agency (i.e., liberal freedom).

Project →

(a) *Intentional self*—in the nondominant position is either low or severely lacking in a sense of agency. Children and adults in the nondominant position perceive themselves as pawns in an arbitrary system which is governed by the hegemony of the dominant class's intentions.

(b) *Individuation*—is lacking in any sustained sense. People in the nondominant social position do not experience that they can have a place in the world carved out by their intentions. They have a sense of being pawns in a larger social system.

(c) *Change*—in a nondominant social position is experienced as disorderly, arbitrary, and unpredictable. Since the larger social totalities that this class is embedded in do not respond to its intentions, change is considered luck rather than the predictable outcome of concerted actions governed by one's intentions.

(d) *Conscious*—actions are not integrated into the larger cultural project. In the nondominant formative position, there is a failure in consciousness. Sartre would call this class a class "in itself" but not "for itself."

(e) *Utopian*—images are muted by the failure of this class to believe that consolidated social action can possibly yield to one's intentions. It is a condition of "bad faith," in Sartre's terms.

(f) *Freedom*—if it is perceived at all, is experienced in the microcosm (i.e., family, peers) but absent in a sense of accomplished within the larger cultural project. This absence of freedom is a condition of what Marx called "alienation."

TABLE 5. Hegemonic Stability (Transformative Dynamics)

Cultural Form of the Nondominant Class		Cultural Form of the Dominant Class
Habitus		Habitus
(a) *Social self*—a sense of solidarity because of common fate and shared oppression. A "need for roots" and a time when oppression was consolidating is expressed, helping to formulate a sense of identity	⟶	(a) *Social self*—the solidarity experienced as being part of a dominant class becomes questionable and there are ambiguous areas which lead to a less stable sense of identity . . .
(b) *Identity*—is accomplished with retrieval of a past history in order to restore a sense of continuity with the present. The authenticity of ancestors and a shared history allows a sense of order	⟷	(b) *Identity*—in which the ties with the past are loosened and a certain amount of instability ensues, which challenges the sense of order . . .
(c) *Order*—which is countercultural and not centered in the history of the dominant group. The ordered legitimacy of schooling and state institutions is questioned as to its ultimacy and importance for the functioning of one's own particular historical group.	⟷	(c) *Order*—which no longer has the uncontested backing and legitimation coming from such ideological sectors as the school, state, and the nondominant class. This lack of an "ordered sense" frequently leads to violence in order to restore the "law and order" of the glorious past.
(d) *Preconscious*—trust in the possibility of a social order geared to the authentic needs of one's group is frequently embedded in a religious perspective, which allows groups to trust in a pilgrimage toward new social forms (e.g., Exodus myth).	⟷	(d) *Preconscious*—trust in the solidarity and naturalness of the present social order, making unquestionableness of the past questionable and leading to a breakdown in the ideological . . .
(e) *Ideological*—order of the dominant group is questioned and rebelled against. The hegemony of the present order is relativized and allows for the rehearsal of emergent social forms.	⟶	(e) *Ideological*—order which either demands a hysterical consistency with the status quo of the past or gradual reforms that allow the integrity of the present order to be retained.
(f) *Determinism*—as to one's place in the present social order is no longer seen as natural and inevitable. Nondominant group begins to see the present dominance of other social groups as neither inevitable nor desirable.	⟷	(f) *Determinism*—the natural entitlement no longer is seen as predictable and solidified, which either leads to a consolidation of the past through some power play or some attempts at change or accommodation.

INTEGRATED ⟵

ALIENATED ⟵

Project →

(a) *Intentional self*—the conception of self in a transformative situation is seen in a dawning sense of agency. The sense that oneself and one's group can create a history rather than being submerged in another's.

(b) *Individuation*—is seen in the experience of a unique sense of identity (i.e., "Black Is Beautiful") with its own unique and idiosyncratic advantages which must be "experienced" and "recognized."

(c) *Change*—in a transformative condition is now seen as a real historical possibility dependent on the ability of one's group to organize (i.e., solidarity) the direction of that change. The hegemony of the dominant group is now seen as systematic rather than arbitrary and, therefore, open to question.

(d) *Conscious*—actions are now organized around the symbols of projected emergent forms. In a transformative situation there is "consciousness raising" and a dawning sense that we are a class—not only a class "in itself" but "for itself."

(e) *Utopian*—images become rich and profuse. When religious symbols are used, the symbols are prophetic in nature. The utopian images ordinarily reject the legitimacy of the present order and project images of emergent forms (e.g., image of the promised land).

(f) *Freedom*—is experienced dialectically. First in the intention of rejecting the absolute ultimacy of the formerly dominant regime and, second, in the sense that by our own agency we can create a new history which is moving away from the oppressive forms of the present.

Project →

(a) *Intentional self*—the natural sense of agency becomes precarious when a dominant class appears to be losing its dominant footing.

(b) *Individuation*—of dominant members becomes curtailed because the scope of the dominant culture's project is partially walled off by the emerging sense of agency of the nondominant class.

(c) *Change*—occurs when the dominant position is questioned and begins to appear disorderly and unpredictable. The loss of agency within the dominant cultural hegemony makes change appear arbitrary and too much in control of anarchic forces (usually another word for the "emergent opposition of the nondominant class").

(d) *Conscious*—actions become hypervigilant, having lost the absolute sense that the cultural project subscribed to by the dominant class is shared even by those who do not share in the dominant hegemony (i.e., nondominant class).

(e) *Utopian*—images become "negatively transcendent" (how do we get back to the good old days?). The more liberal utopians concede that reform and development of the dominant cultural project is imminent and a sense of gradualism will best accomplish the necessary changes.

(f) *Freedom*—which particularly came from the sense of agency over nondominant classes becomes curtailed. This unlimited freedom is suppressed by the emergent opposition of the nondominant classes presenting alternative projects. "New freedom" is possible if there is an authentic identification with the opposition of the nondominant class.

structure, I must insist that the reader understand the provisional nature of these outlines. Class dynamics must be seen in relational terms of power and legitimation; it would be naive to see my tables as suggesting a simple two-class analysis. In a complex society like our own, it is necessary to comprehend that individuals and groups do not occupy absolute positions and that there are "contested terrains" (Edwards, 1979). This chapter was not meant to exhaust these subtleties but to orient the reader toward the importance of considering the dynamics of class in the development of the personal world of specific cultural forms.

CLASS, CULTURAL FORM, AND ETHNICITY

With the exception of the historic English working class, it is difficult to pinpoint a class *per se* as a specific "cultural form." I would contend that the personal world as a cultural form is experienced through ethnicity. In other words, people perceive themselves ethnically before there is a consciousness of class location; that is, the premptive experience of the personal world is culture rather than class. One of the most important features of contemporary Marxist analysis is the attempt to relate culture dialectically to class (see Gramsci, 1971; Williams, 1976a). It is here that I take a point of departure. I will try to relate the personal world as a cultural form to the larger structural totality of economic class analysis. Up to now, I have developed the notion of the personal world as a totality, that is, the relational (I and you). I also developed the idea of the personal as a cultural form; that is, we are not persons in relation in the abstract but real histories. Our personal world is a lived history of specific cultural forms or cultures (e.g., Pole, Swede, black, Puerto Rican, Native American, etc.). In this chapter we are embedding the personal world in the larger structural totalities of culture, class, gender, and age. I commence by following Williams (1976a), who emphasizes the "intentional nature" of a class structure analysis. The previous chapters have prepared the way for considering the intentional nature of class. Williams (1976a) sees an emphasis on "intention" as important for the following reasons:

> For while it is true that any society is a complex whole of such practices, it is also true that any society has a specific organization, a specific structure and that the principles of this organization and structure can be seen as directly related to certain social intentions, which in all our experience have been the rule of a particular class. . . . But in many areas of social and political thought, certain kinds of ratifying theory, certain kinds of law, certain kinds of institutions, which after all in Marx's original formulations were very much part of the super-structure—in all that kind of social apparatus and in a decisive area of political and ideological activity and construction if we fail to see a superstructural element we fail to recognize reality at all. These laws, constitutions, theories, ideologies, which are claimed as natural, or as having universal validity and significance, simply have been expressing and ratifying the domination of a particular class. (p. 204)

Williams (1976a) further suggests that the notion of a social totality should be combined with that of the crucial neo-Marxist concept of hegemony. Here the thought of Gramsci (1971) assumes prominent significance (Williams, 1976a). For Gramsci, *hegemony* refers to a form of ideological control in which dominant social practices, beliefs, and values are reproduced and disseminated through a range of institutions such as schools, family, mass media, and so on. Hegemony assumes the existence of a totality that saturates the society to such an extent as even to constitute the limit of common sense for most people under its sway. The important feature that Gramsci puts forward in the concept of hegemony is the emphasis on the reality of *domination*. Williams (1976a) suggests that in any society, in any particular period, there is a central system of practices, meanings, and values that can clearly be seen to be dominant and powerful. Succinctly:

> It is a whole body of practices and expectations; our assignments of energy, our ordinary understanding of the nature of man and of his world. It is a set of meanings and values which as they are experienced as practices appear as reciprocally confirming. It thus constitutes a sense of reality for most people in the society, a sense of the absolute because experienced as reality beyond which it is difficult for most members of the society to move, in most areas of their lives. (p. 205)

Hegemony, because of pervasiveness, relates to all major spheres of social existence; it is convenient here to consider four major ideological realms: (1) the economic realm, which encompasses the ideologies of production, exchanges, distribution, and so on; (2) the cultural realm, encompassing the ideologies of culture, values, mass media, and so on; (3) the political realm, encompassing ideologies of the state, legal–judicial system, police, military, and so on; and (4) the social realm, encompassing ideologies of the private sphere, family, education, social groups, and so on (Kellner, 1978). Psychology and the "personal world as a cultural form" is part of this last realm. One therefore must see this personal sphere as embedded in these larger totalities.

The question that presents itself immediately is: How can this personal world have an integrity if it is totally embedded in the hegemony of the dominant culture? This is an important issue and demands some subtlety in analysis. If the powerful legitimations of modern capitalism exert themselves with such a dominating force in all ideological realms, it would seem that there is no room for any independence from this pervasive consciousness. To say that our personal worlds are not powerfully determined under the hegemony of capitalism would be patently naive. Left at this point, however, we are locked into a total determinism of reproduction ("Whatever has been will be"). Williams (1976a) introduces the notion of oppositional forms to the dominant ethos as a way of handling this problem:

> We have to think again about the sources of that which is not corporate of those practices, experiences, meanings, values, which are not part of the effective dominant

culture. We can express this in two ways. There is clearly something that we can call alternative to the effective dominant culture and there is something we can call oppositional in a true sense. The degree of existence of these alternative and oppositional forms is itself a matter of constant historical variation in real circumstances. (p. 206)

Oppositional Forms as "Residual" or "Emergent"

Williams (1976a) identifies two oppositional forms within the hegemony of the dominant culture. A "residual" is in essence a carryover from a previous historical period. The presence of a residual form is indicative of some experiences, meanings, and values that cannot be expressed or verified in terms of the dominant culture but are, nevertheless, lived on the basis of the residue of some previous social formation. Certain forms of bartering practiced in certain rural areas offer one case in point.

The second is an "emergent" form in opposition to the dominant culture. Emergent forms assume that new meanings and values, new practices, new significances and experiences are continually being created (Williams, 1976a). Williams makes an important distinction between what he calls "alternative" and "oppositional" forms:

> There is a simple theoretical distinction between alternative and oppositional, that is to say, between someone who simply finds a different way to live and wishes to be left alone with it, and someone who finds a different way to live and wants to change society in its light. This is usually the difference between individual and small-group solutions which properly speaking are ultimately revolutionary practices. But it is often a narrow line, in reality, between alternative and oppositional. (pp. 206–207)

Brown (1973) identifies these oppositional forms to the presently constituted "bourgeois hegemony." They are seen in the development of revolutionary movements in both the third world and among the colonized minorities of western metropolitan countries. The oppositional groups to the dominant culture are characterized as "youth culture," "feminist" movements in the west, "black power" and black nationalism (Brown, 1973). As emergent cultural forms, they provide the basis for new identities and, therefore, new cultural projects that challenge the hegemony of the status quo. Combined, these cultural movements are potentially the basis for an alternative cultural identity. There must, therefore, be a sensitivity to emergent cultural forms or countercultural oppositions that are not simply considered as deviant. A sensitivity to all "liberation movements" on their own terms is an important prerequisite. This does not exclude a critical analysis; rather, the critical analysis is based on the ultimate viability of cultural forms that potentially challenge the destructive aspects of cultural forms developed under monopoly capitalism. As social interpreters or hermeneuts, we do not rush in with the view of treating oppositional groups as "deviant." Even when sympathetic, one does not end up by "blaming the victim," which is a common practice of mainstream psychology toward vulnerable cultural groups (Ryan, 1971; Glasgow, 1980). Indeed, it is the task of a critical psychology to identify the hegemonic structural factors in the organization of our society which create systemic oppression for specific cultural groups (e.g. blacks, Native Americans, women, children, etc.). Let me now turn to some *specific* examples of this type of analysis by specifying how they can be related to gender, age, and racial and ethnic inequality.

A Case Study of Ethnicity and Class Dynamics

The study that I am now about to interpret combines class and ethnicity within its overall analysis. The work is Glasgow's (1980) *The Black Underclass*, subtitled *Poverty, Unemployment and the Entrapment of Ghetto Youth*. This is the study of the young blacks in the Watts ghetto and specifically involves the Watts uprising. Glasgow (1980) calls them the "underclass" because they are an identifiable core of black youth that persist from one generation to another in a condition of chronic long-term unemployment and who remain persistently poor and immobile in the economic structure.

Racism, it would appear, is as old as humankind itself. All major social systems in history have suffered from it in one form or another. With this in mind, it may be said that monopoly capitalism does not have a monopoly on aberrant race relations. It does, however, utilize racial differences for its own peculiar designs. Racial differences are a convenient marking for class hierarchies. The mark of the stranger has always been a convenient justification for oppression and exploitation. In a capitalist class-structured society, racial differences allow the possibility of mapping social class differences along racial lines.

No one here is denying that there are differences in cultural forms that proceed along the lines of racial characteristics. This is not an issue here. What will be questioned is (as in the case of gender relations) the pervasive use of organic metaphors that have been used to explain racial differences. In addition, the organic metaphor will be questioned in the case of racial differences, insofar as it seems systematically to legitimate racial exploitation along class lines. Our contention is that racial groups constitute cultural forms that go into the makeup of unique personal worlds. The ambiguity of class boundaries in advanced capitalist society (see Wright, 1976) demands a treatment of race and class of some subtlety. Racial dynamics interact within a class structure in quite an intricate fashion. Even in racial groups whose class origin is similar (i.e., working class), racial division, as with gender division, consolidates the dominant hegemony (Willis, 1977). The divisiveness of cultural forms along racial lines has a considerable amount of psychological dynamics (Willis, 1977). It is here that a more subtle treatment of the relational dynamics of class and race becomes necessary. In an advanced capitalist society one can observe sociological distinctions between middle class, working class, and lower class subcultures (see Piore, 1973). The complex racial dynamics within a class-structured society create different mobility chains (Piore, 1973). Racial dynamics in the lower mobility chains (i.e., working class, lower class) divides and conquers any intimations of solidarity (Willis, 1977):

> It also provides an evident underclass which is more heavily exploited than the white working class, and is therefore indirectly and partially exploited by the working class itself (at least lessening their own exploitation); it also provides an ideological object for feelings about the degeneracy of others and the superiority of the self (thus reinforcing the dominant ideological terms which make the comparison possible). Racism therefore divides the working class both materially and ideologically—Racism must be understood with respect more to the complex social definition of labour power under capitalism than to any pure and inevitable ethnic hostility. (pp. 152–153)

Although there are many important insights in this study, the part that I will accent is the dynamics of the personal world under a formative dynamics, already discussed in Table 4, and the nascent transformation of that world (transformative dynamics, see Tables 5 and 7), which ensued with the initial eruption of the Watts riots in Los Angeles in the late 1960s. The importance of this study here is its exemplifying how the cultural world of a particular racial group transforms itself quite dramatically when the class structure which entraps them is rebelled against. The following is my own interpretation of Glasgow (1980) within the scheme formation–transformation dialectic that has been outlined previously in Tables 4 and 5. Tables 6 and 7 exemplify the structural and personal dynamics of this group of youthful blacks, who are entrapped in the lowest tier of the class system. Table 6 outlines the formative dynamics that were dominant before the Watts riot. Glasgow (1980) indicates that it is necessary to have a radically different approach to social inquiry in order to understand the complex and dramatic alteration of the social world that accompanied and followed the Watts uprising. The first violence is a striking out against the institutions and symbols representing oppression and exploitation.

> The ghetto seethed, tensions mounted, and the police—the hated symbol of black containment in enclaves north and south—finally provided the catalyst that set in motion one of the most ravaging outbursts of blacks in the history of this nation. Their rage was directed at white society's structure, its repressive institutions and their symbols of exploitation in the ghetto: the chain stores, the oligopolies that control the distribution of goods; the lenders, those who hold the indebtedness of the ghetto bound; the absentee landlords; and the agents who control the underclass while safeguarding the rights of those who exploit. The explosion was intended to get even with those whom poor blacks hate and resent, the people who profit from the powerless condition of the poor and the entrapped; it also represented the hope of making some change in their living conditions. (p. 106)

The dramatic shift that occurred after the initial riot cannot be explained in any linear interpretative system. The personal world obviously can change in some quantum fashion when the oppression of the class and racial dynamics are challenged directly.

> A different approach is needed to answer the question of what propelled a young man in 1965 to throw a blazing bottle of gasoline into a store front in the late afternoon when only hours earlier he was with a friend at the beach making love. And it is more important to ascertain what encouraged the same young man to move from violence and destruction to activities for community improvement all within the short period of six months. In the end, we must ask what were the social mechanisms making such dramatic movement possible—from violence to constructive involvement, from gang hooliganism to responsible participation. (Glasgow, 1980, p. 107)

The first attempt at launching an authentic project is through active resistance toward the entrapping social structures. Violence is the initial volley. As one participant put it:

> The brothers got together. I was hollering, "Burn, Baby, Burn." Lou Rawls expressed it well: "Tear it up, blow it up, Tobacco Road, and build it up again." This was the feeling I had. I wanted to destroy and rebuild. (Glasgow, 1980, p. 112)

For the next year this initial resistance consolidated into a sustained community project signaled by the development of a community action group of youth called The Sons of Watts. This community mobilization group were the youth who were formerly "hanging out in the streets." An organization called SWIA (Sons of Watts Improvement Assocation) was started. This community action group was initiated at the grass-roots level and did street cleaning, policing, job training, personal growth programs, and so on. The group was a community expression of black unity: "Everybody was sticking together. Everyone was calling each other 'Brother.' People were more civil and there were no more fights between the brothers. Everyone spoke to each other" (Glasgow, 1980, p. 117).

This sense of solidarity and participation created a sense of agency which issued from the development of a grass-roots community project. The sense of a project moved from initial resistance against the oppressive forces of the ghetto to a belief that with organization and solidarity their lives could possibly change for the better. A positive sense of self-worth followed without the aid of any psychological services.

> The thing that got me was the role I played in maintaining order in our community. I became dedicated as I saw everyone saying "welcome to Watts." No one was ashamed (of the rebellion), they were proud. It made me proud of myself, expecially to my brothers and sisters.
>
> I saw the power we had during the festival and the people's respect for us. When I decided to come (to the festival), at first I saw it as a good game. Then we talked about the Sons of Watts; it sounded like it could be something. Then I became a Son of

TABLE 6. Schematic Interpretation of Watts Youth before the Ghetto Riot—Formative Dynamics of Underclass Youth Entrapment before Riot

Habitus		Project
(a) *Social self*—in underclass entrapment, the social self accepts the negative identity of the dominant class. Where there is some positive self-conceptions, it is frequently coalesced around the ability to survive in conditions of dire and chronic poverty.	⟷	(a) *Intentional self*—is totally occupied with the survival demands of the ghetto life. Although there is resistance and rebellion, it is not directed at the system and is not felt possible to be effective. There is a sense of being a pawn in the larger economic system, which defies attempts at agency.
(b) *Identity*—is achieved by the repetitive quality of survival conditions in chronic unemployment. The repetitive quality of these negative survival conditions engenders a sense of negative self-worth. Survival culture is reactive in origin but not a passive adaptation to encapsulation. Activity sustaining an identity can be extremely resistant, but completely in response to rejection and the destructive aspects of ghetto life.	⟶	(b) *Individuation*—is stifled by being locked into a system that hampers movement. A certain cultural style is present, but it is an individuation that has been totally styled by a "history of oppression and exploitation." Within the ghetto a sense of individuation is achieved by proving you can "take a piece out of the system." Thus "conspicuous consumption" becomes one expression of individuation in the ghetto. This consumption usually mimics the consumption patterns of the dominant culture.
(c) *Order*—is achieved by the legitimacy of welfare and state intervention under liberal capitalist hegemony. This minimum order in an otherwise chaotic life is achieved bureaucratically through the facilities of welfare programs, which are permanent when high unemployment is acceptable within the state. There is also a community of survival, which functions at a day-to-day level but at the mercy of police intervention. The presence of the police represents the organized violence of the state maintaining order in ghetto life.	⟶	(c) *Change*—in the ghetto is experienced negatively. In the lowest tier of the economic system, change is usually for the worse. You are more likely to be losing a job than finding one, going to prison rather than coming out, losing a friend through self-destruction rather than celebrating his or her good fortune. Change is therefore experienced as a "downward spiral" whose cycle can only be broken by steady employment, "which not only earns money, but also provides opportunities to invest in a future"; this has much to do with building one's sense of self-worth. Thus, "I need a *job*, a job where it's at" (Glasgow, 1980, pp. 71–72).

(d) *Preconscious*—solidification of ghetto entrapment is engendered by a generational cycle of chronic unemployment. The situation of a crisis state and its survival measures become "natural" and "everyday" rather than temporary feelings of lack of security that most people feel.

(d) *Conscious*—actions leading to goal-directed projects are absent. The ethics of survival lead to a hand-to-mouth mentality, which curtails the projection of sustained and long-term problem solving. Glasgow notes that institutionalized racism—characterized as ghetto residence, inner-city educational institutions, police arrests, and so on—does not produce lower motivation; it destroys motivation.

(e) *Ideological*—pursuits of the dominant class are accepted by ghetto youth. Being affected by the mass media, they accept the desirability of consumption and flaunt it when possible. Most of the time, however, they eventually accept that that they will not get "a piece of the pie." This is ideologically backed up by the ghetto education, which usually ends in failure. Contrary to popular belief, ghetto youth aspire to graduate from high school; their educational aspirations are dampened as they proceed through a school system that perpetuates a sense of permanent underachievement.

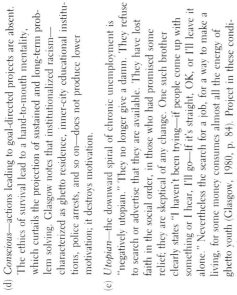

(e) *Utopian*—the downward spiral of chronic unemployment is "negatively utopian." They no longer give a damn. They refuse to search or advertise that they are available. They have lost faith in the social order, in those who had promised some relief; they are skeptical of any change. One such brother clearly states "I haven't been trying—if people come up with something or I hear, I'll go—If it's straight, OK, or I'll leave it alone." Nevertheless the search for a job, for a way to make a living, for some money consumes almost all the energy of ghetto youth (Glasgow, 1980, p. 84). Project in these conditions is limited to survival maneuvers. A large societal project is totally entrapped under conditions of chronic unemployment.

(f) *Determinism*—is felt as the changelessness of chronic unemployment. After these young men learn the educational system, we have the picture of a group of youth who have no educational credentials and little vocational training in the skills necessary for the swiftly changing job market—a group of men totally at the whims of periodic employment.

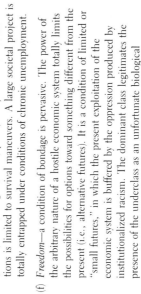

(f) *Freedom*—a condition of bondage is pervasive. The power of the arbitrary nature of a hostile economic system totally limits the possibilities for options toward something different from the present (i.e., alternative futures). It is a condition of limited or "small futures," in which the present exploitation of the economic system is buffered by the oppression produced by institutionalized racism. The dominant class legitimates the presence of the underclass as an unfortunate biological occurrence.

TABLE 7. Schematic Interpretation of Watts Youth after the Ghetto Riot—Transformative Dynamics of Underclass Youth's Attempts at Transcending Entrapment

Habitus		Project
(a) *Social self*—there is a rejection of the "white man's" definition of their identity. This is initially seen in the resistance to police arrests and rejection of the "law and order" forces as enforcers of the white man's order.		(a) *Intentional self*—is seen not only in resistance to the projects of "whitey" but also in the "new sense of agency" achieved through the development of grass-roots organizations. There is a dawning sense of the possibility of being an "origin" rather than a pawn in an alien system.
(b) *Identity*—there is a rejection of the negative identity produced by survival tactics of the ghetto. A new community solidarity with the positive sense of self-worth ensuing allows for a feeling of "blackness" that has positive connotations. There is a strong sense of black solidarity.		(b) *Individuation*—is achieved by a new sense of societal project and a more positive sense of identity. "Black Is Beautiful" means not only a rejection of whites but a sense of uniqueness that there is in being black. This individuation is not individualism, it is a unique "we" consciousness.
(c) *Order*—the hegemony of the "order of the welfare system" is broken by the development of community organizations run by blacks to deal with black needs. There is a rejection of "state order," signified by the attitude toward the police, which is negative; also a sense that the community can provide its own indigenous order through grass-roots organization work.		(c) *Change*—the sense of altering the downward spiral of ghetto life after the riot is very strong. The belief in change and in that sense upward transcendence unlocks a flurry of activity which exemplifies a sense of exhilaration. "We're going to change things now" is a prominent slogan.

(d) *Preconscious*—solidification of "ghetto entrapment" is questioned with the onset of the riot. The riot, it may be said, is an awakening from the slumber and narcosis of ghetto entrapment. In very specific ways, the spontaneity of the uprising is nevertheless accompanied by a clear sense of what establishments are the exploiters in the ghetto. This preconscious understanding marks out specific establishments for looting and destruction, while others go unharmed.

(d) *Conscious*—actions became a dominant motif in the post-Watts uprising. A heightened awareness of the possibilities of new problem-solving actions through community organization is present. There is a great deal of planning for a different future where before there was none.

(e) *Ideological*—rejection of the status quo and the state structures are accompanied by an increasing identification of a larger black consciousness and a longer black history that spans their own individual histories.

(e) *Utopian symbols*—become extremely important. The possibility of having a different black future is celebrated in symbols, dance, and theater. The new community awareness and solidarity are signified in symbols that appear to challenge the downward spiral of former ghetto life.

(f) *Determinism*—the hegemony of the past can be changed. "Whitey" is going to have to do some moving now. The strong feeling of the weight of an oppressive past is challenged.

(f) *Freedom*—is seen by the belief that the future will not be a repetition of the past. This sense of new project is the very essence of a notion of freedom.

Watts. It's a good thing. Let me tell you, we are a group of unique people; had a chance of being the world's greatest pimps and gangsters. We acted bad; now we can be respected. We burned four million dollars worth of property; we have power and should be respected. We're trying to do something different now. (Glasgow, 1980, pp. 127–128)

The tragedy of all this is seen in the ultimate fragility of small-group movements that challenge the governing hegemony of capital. Watts is a case in point. Glasgow (1980) brings his sensitive analysis to a sobering conclusion.

When the real issues of productive functioning are examined, the most obvious fact is that these men continued to be jobless, poor, and occupationally obsolescent. Despite their efforts at self-help, no stable economic future was in sight. For the most part, the income-producing programs did not materialize, and although their job-hunting and market-analysis skills became a little more sophisticated, these improvements did not pay off in increased employment because the opportunity was not there. The self-help community-service-oriented programs, although very few, nevertheless continued to be the most lasting functional part of their activity. . . . Much of the early enthusiasm for the Sons had worn off by 1975. The equation of oppressive poverty had not been broken, and instead increasingly broken was the will of many of the men. Opportunities for stable income-producing jobs and the avenues to upward mobility remained closed. (p. 149)

This indigenous group (Sons of Watts), while it lasted, provided opportunities for underclass men to express themselves and to countenance a sense of agency and project. This phase is schematically illustrated in Table 7 as the transformative dynamics.

The Glasgow study on the black underclass is a specific example of the complex interplay between class dynamics, ethnicity, and the development of a personal world. It is part of a genre of literature that developed in the wake of the black movement in the 1960s and an in-depth examination of the black experience, life-styles, community organizing, interpersonal and family relationships, and so on. I view it as a critical interpretation of the black experience and, at the same time, a study of the relationship of class and ethnicity. Its critical intent is expressed in its explicit departure from mainstream social sciences dealing with ethnicity and more specifically, in this case, black ethnicity.

For example, new importance was given to ethnicity as a factor in black oppression and to liberation strategies as against class factors; positive ethnocentrism was elevated in contrast with the former deviant subculture thesis. Most important was the reaffirmation of the black family structure, and one's function therein, as a prime contributor to blacks' survival. This emphasis replaced the persistent notion promulgated by mainstream social scientists that the Negro family was in disarray and that, in fact, it was the primary source of the second-class position of blacks and the conveyor of a disadvantaged culture.

Instead, this study shows how a specific cultural form develops through a unique sense of historicity (ethnicity) but embeds that history within the complex confines of a class analysis. Within this complex dialectic, the personal world and its vicissitudes are explained with a disciplined and critical sensitivity.

The previous example was given to illustrate the relationship between class structure ethnicity and the dynamics of the personal world of specific cultural forms expressed in ethnicity. I would now like to turn to the role that gender plays in a critically interpretative enterprise.

GENDER AND DOMINATION

Changing cultural standards do not alter the fact that sex differentiation constitutes almost a bifurcation in status and roles in almost all cultures from the very beginning of the socialization process. All known human populations appear to honor this dichotomy, and all cultures distinguish males and females on the basis of primary and secondary sex characteristics (Ausubel *et al.*, 1980). By gender, I refer to role attribution between the sexes based on both primary and secondary sex characteristics. The category differentiation based on gender is universal, and this differentiation transcends the category systems of class and ethnicity. In other words, there are women and men in all races and classes.

The notion of gender domination signifies that there are power differences that accrue from the attributions of gender. In most cultures, and in ours specifically, men have historically dominated women in terms of the power relations operating in society. This is referred to as *patriarchy*. Patriarchy signifies that men have premptive power over women at both the economic and cultural symbolic levels. In other words, women occupy an inferior and dependent status that is culturally sanctioned. Historically, this predates capitalism and can also be seen in structures that attempt to transcend capitalism (e.g., eastern European socialism). It appears to be the most universal form of oppression and domination.

I intend to restrict my comments here to a discussion of patriarchy under capitalism. My discussion will suffer from its brevity, but in a critical psychology the issue of gender domination cannot be sidestepped. Let me commence by noting the sustaining power of the gender system and male dominance within it, which has been engendered by the pervasive use of the biological or organic metaphor in the historical explanation of sexual differences. The organic metaphor, as we have already suggested in Chapter 1, is a conservative metaphor for the maintenance of social structures being interpreted. From Aristotle through Jung and Freud, the use of biological metaphors has put woman in the category of "incomplete man" (Whitteck, 1976). At the beginning of the 20th century, biological theories of sex differences were systematically ideological in nature; helping to refute from a so-called scientific perspective the claim of feminists at that time to equal rights for women (Hall, 1976). Even today, biological metaphors of gender differences are used by conservative scholars to sustain the rigid legitimation of fixed roles (Markovic, 1976; Wine, 1980). Now, however, these perspectives are less blatantly sexist and their edges are blunted by technical issues (Hall, 1976). A consideration of women as agents in the appropriation of a personal world is absent from most conceptions of human sexuality where metaphors other than the organic are used (e.g., mechanical metaphors are frequently utilized in obstetrics). Notwithstanding the caveat above, it may be safely said that in terms of gender, biological metaphors predominate (see Chodorow, 1978; Dickason, 1976; Eichler, 1980; Gould, 1976; Hall, 1976; Oakley, 1979; Marcovic, 1976; Whitteck, 1976). Chodorow (1978), in criticizing biological metaphors, reiterates the importance of the personal metaphor that this book is being espoused for an understanding and interpretation of the gender system.

> The biological argument for women's mothering is based on facts that derive not from our biological knowledge, but from our definition of the natural situation as this grows out of our participation in certain social arrangements. That women have the extensive and nearly exclusive mothering role they have is a product of a social and cultural translation of their childbearing and lactation capacities. It is not guaranteed or entailed by these capacities themselves. (p. 30)

The personal world, then, molds the contours of capacities and gives them specific use values. In the structural system of capital, the agency of the personal world is stunted and blocked more for women than for men, regardless of class. In that sense the gender system has its own peculiar forms of oppression, which are projected into the larger class system and conversely affected by that system of commodity production. Thus, the oppression produced by the domination existing within the gender system crosses all classes (i.e., male domination) but affects women and men differently in terms of their relations in the larger class structure.

Patriarchy and the Domination of Women's Projects

I have indicated that the focal point of human freedom is in the possibility of experiencing agency in the creation and interpretation of one's world (i.e., project). In this sense, it may be safe to say that women of all classes and races have been denied this sense of agency historically. Patriarchy, as I am using the term, is the systematic masking of the projective character of women's action by men. In most cases, this systematic oppression of men over women is unconscious (i.e., habitual and part of the cultural habitus). This domination of women becomes conscious and intentional historically when women attempt to assert themselves against the dominance of the male project. The domination of women historically takes the form of male condescension or outright violence. Let me briefly schematize structural dimensions of the oppression engendered by gender domination without considering its specific historical contours within class and racial structures. When patriarchy is stabilized, one can say that the structural conditions of oppression are "hegemonic" in the Gramcian sense of this term.

Gender, Class, and the Reproduction of Social Relations under Capitalism

The reader should understand that the schema above (Table 8) should be taken in with a certain cynicism. If left without further comment, this schematization would be ahistorical and would thus defeat the historical sensitivity that I am inviting the reader to consider. This is important, because what I am now about to stress is the historical specificity of patriarchial structures within a capitalist mode of production. This means that male dominance takes historically specific contours within different economic systems. For example, gender domination, until recently, was accomplished through the structure of the nuclear family in modern capitalism. This, as already stated, is done by the separation of the world of work (i.e., production) from the world of the family (i.e., reproduction). In connecting these worlds, Wolff (1976) points out that the most important single act of social production is reproduction, which is not only the conception of children but the caring and raising of them to maturity. But this is done differently in different historical periods. It is only with the beginning of industrialized societies that the separation of work life and family has become prominent and apparently necessary (Kamerman, 1979). The separation of the world of work and family has peculiar consequences for the sexes within capitalist production (Chodorow, 1978; Marcovic, 1976). "Parenting" within the social organization of the nuclear family produces sexual inequality, not simply role differentiation (Chodorow, 1978). This is accomplished by first locating women and men differently within the organization of production (i.e., economic system). Men are, in essence, equated with the public sphere while women are equated with the private (Wolff, 1976). If privacy were the salient characteristic of the personal world, women would be considered more personal and men more impersonal. The position I am venturing rejects a notion of the personal world as private and individual. The differentiation along these lines reflects poorly on both sexes. Hegel was the first modern thinker to systematize the distinction between the public and private spheres (Gould, 1976). He identified the family with the private domain of love and immediacy, whose main function is that of the socialization of children. This was to be contrasted with "civil society," which is based on need and labor rather than sentiment. Gould (1976) sees this distinction as a major mystification, arguing that the family is pervaded throughout by bourgeois or civil society, and that the concrete nature of the family as a social and political institution is concealed under capitalism. In modern monopoly capitalism, the family is a major consuming unit and the target of advertisement unknown in previous historical periods (Ewen, 1976; Gould, 1976; Marcovic, 1976; Sullivan, 1980b).

TABLE 8. Hegemonic Patriarchy

Male Dominance — Habitus		Woman Oppression — Habitus
(a) *Social self*—a solidarity between men across classes that to be a man is to be identified with a superior gender. Male dominance indicates that the part of one's overall personal identity which is gender-related is conceived of in positive terms.	↔	(a) *Social self*—the social self is permeated with the sense of inferiority of the women's sex from the cultural surroundings. This is stabilized by the sense that this "inferiority" is biologically based and invariant historically.
(b) *Identity*—male identity is integrated into the wider culture's valuing of the male sex. In extremely macho cultures, this can be seen in the sadness of the family when the first child born is a female.	↔	(b) *Identity*—a positive sense of the self is not buttressed by the surrounding culture. Identity is achieved in a more derivative manner—females get to perform roles that are respected by male-dominance norms.
(c) *Order*—is achieved by the understanding that cultural structures which make female oppression unconscious (e.g., from the earliest socialization practices, both mothers and fathers treat females as being more dependent and vulnerable in relation to male children).	↔	(c) *Order*—is achieved within the understanding that females will be dominated by males and thus be protected by the "stronger sex." This notion of the stronger sex concedes that male might makes right.
(d) *Preconscious*—male dominance is achieved by cultural structures that make female oppression unconscious (e.g., from the earliest socialization practices, both mothers and fathers treat females as being more dependent and vulnerable in relation to male children).	↔	(d) *Preconscious*—a habitual status of dependency and vulnerability for female children allows for an unconscious appropriation of cultural dependency structures that are perceived to be biologically induced rather than culturally learned.
(e) *Ideological*—underpinnings of patriarchy are buttressed by biological metaphors which culturally sanction male dominance because of inherent biological differences where males are superior biologically.	↔	(e) *Ideological*—underpinnings for female oppression are buttressed by biological metaphors that culturally sanction female inferiority because of inherent biological differences where females are inferior biologically.
(f) *Determinism*—male agency and project are believed to be inevitable and natural consequences of biological superiority.	↔	(f) *Determinism*—female oppression by male dominant structures are seen as inevitable and natural because of biological inferiority.

INTEGRATED ←

 ALIENATED ←

Project

Project

(a) *Intentional self*—males perceive themselves to be the focal point of agency in a society under patriarchy. Males perceive that their intentions are preeminent over females because "this is a man's world."

(b) *Individuation*—is the sense that there is a place in the world for me which I, through my agency, bring about. Although class structures make a difference where males are concerned, patriarchy crosses these boundaries in distinct ways.

(c) *Change*—the ability to reorder the world is understood to be relatively possible for males even where class domination is operating. So, even under conditions of class domination, lower-class males have a greater sense of agency than their female counterparts.

(d) *Conscious*—actions are governed by directives of the larger patriarchial project of men over women. Men see themselves as more goal-oriented than women, exclusive of class structures and class dominance.

(e) *Utopian*—images of change are imaged as being carried out by solidarity among men within a background of women's support. Male utopian images frequently emphasize control and power over participation and community.

(f) *Freedom*—is sensed under patriarchy to be the result of male initiative or agency. Even under conditions of class dominance, lower-class males have a greater sense of agency than their female counterparts.

(a) *Intentional self*—under conditions of oppression within patriarchal structures, women are either low or severely lacking in a sense of agency. They perceive their intentions where operating to be subordinate to their male counterparts.

(b) *Individuation*—under patriarchy is muted and eclipsed because the nondominant social position of women makes it appear inevitable that a world carved out by their intentions is impossible to achieve in an order of male dominance.

(c) *Change*—for women under patriarchy is sensed to be possible as a result of male agency. Thus, women under stabilized patriarchy see change as possible only as the result of male initiative.

(d) *Conscious*—actions for women are habitaully oriented to the projects of men under patriarchy. Supposedly behind every male initiation or project is a woman who supports his intentions without reciprocity.

(e) *Utopian*—images are muted and fragmented by the failure of women, under patriarchy, to believe that consolidated social actions could be launched by women as a cultural force. Frequently, this fragmentation of images is exacerbated by competition between women for men's favor.

(f) *Freedom*—for women under patriarchy is perceived as coming through the initiation of male agency. One could label this "spurious freedom of alienation," since it is not tied to the direct agency of women.

Insofar as woman in the family is exploited (by performing unpaid labor) and oppressed (by having to take money from her husband), by being commodified and
manipulated by advertisements and the media, by being coerced to raise the children
by herself and in "appropriate" ways, the political emancipation of the woman, as well
as the assurance of equal opportunity for her, reveals itself to be inadequate. Mini-
mally, we might say that concrete emancipation requires the overcoming of the
dependence of the family on capital. (Gould, 1976, p. 36)

The situation reveals its complexity when one understands that gender
ideologies are reproduced within the family structure itself. For example, lower-
class males enforce the ideology of domesticity on women, which is a virulent
sexism. Especially because of the alienating nature of work under capitalism,
men reinforce the sexual division of spheres as a defense against powerlessness in
the labor market (Chodorow, 1978).

It is important to understand the social construction of the world of the
nuclear family for our understanding of the reproduction of the ideology of
capital. Chodorow (1978) points out how this is reinforced by mothering in the
nuclear family. Mothering in the isolated nuclear family of contemporary capitalist society reproduces specific personality characteristics in men that reinforce
and replicate an ideology of male superiority and a concomitant submission to
the requirements of production (Chodorow, 1978). Specifically, this is done by
preparing men for participation in a male-dominant family life and also for their
participation in the capitalist world of work (Chodorow, 1978). When women
enter into the public sector of work, which is done increasingly, this does not
mean relinquishing their home making responsibilities. In the work force, male
domination is seen in economic disparities between men and women and job
opportunity. Hutner, an economist, contends that working women are trapped
in female "occupational ghettos." They receive low-paying clerical salaries in
general, but even in the more prestigious professions (i.e., teaching, liberarian-
ships, nursing, etc.), women are consistently paid lower salaries than men.

The gender system supports the class structure of capital through a complex
process mediated by the modern nuclear family. A weak unit for the sustenance
of its own designs, it nevertheless functions well carrying out the consumer needs
of the corporate system (Sullivan, 1980b). The reproduction of the social relations of production through the reproduction of culture is amplified by the
development of a consumer mentality, in which women and now children in
our culture are expected to excel (Sullivan, 1980b). This role is not, however,
complementary and is only indicative of woman's oppression in the gender
system as well as the class system (Marcovic, 1976).

Private property, in the form of capital, is still everywhere essentially in men's hands.
There is one economic aspect in which women are relatively powerful: that is consumption. However, the distinction between private property with respect to the
means of production and private property with respect to the means of consumption is
very important from the point of view of the distribution of total social power. Those

who dominate production also control the huge advertising machinery for the creation of artificial needs and the demand desirable at the market. Thus, consumers hardly exert a very significant influence on those who control production; consumers are rather manipulated by them. (Marcovic, 1976, pp. 155–156)

Ethnicity, Culture, and Gender Domination

In a short treatment of such a controversial topic, I feel responsible not to spare the reader some of the complexities which brevity may tend to conceal. One of these complexities is the realization that gender structures and patriarchy are also formed differently within the context of culture and ethnicity. For example, male dominance manifests itself differently in North American culture when compared with the culture of Latin America.

It is informative to make a comparison between North America and Latin America in relation to the hegemonic acceptance of class and gender structures. In North America, the economic order is stabilized and hegemonic; that is, there are no revolutionary upheavals that question the fundamental integrity of the economic order. At the same time, the women's movement, that is, the movement which fundamentally questions patriarchy, is quite vocal and active. The exact opposite prevails in Latin America. There are active revolutionary movements throughout Latin America which fundamentally question the continuing stability of the economic order. Most of these movements do not incorporate within themselves a rejection of oppressions which are gender-related and due to the male dominance of patriarchy. I use this example for one reason only, and that is to sensitize the reader to the complex relationship of patriarchial systems to class and ethnic structures. It will be the task of a more mature critical social inquiry to explore and elucidate the complex relationships between these systems of domination.

Patriarchy Challenged—The Women's Movement

Probably the most steady and insistent movement for cultural change in the last 20 years has been the women's movement, challenging cultural stability based on gender-domination of male over female. This movement is quite complex and varied, takes many complex routes, and has many guises. What is at the bottom line is a challenge made by large numbers and diverse groups of women, with the support of a minority of men, to the accepted habitus that men should dominate women as a cultural project. Part of the challenge of this movement has been in the erosion of sterotypical male and female roles. This erosion is exemplified in challenges to role stereotyping of women by men and also catalyzed by the articulation of atypical male roles that seek wider cultural acceptance (e.g., gay liberation). It is seen in the challenge that the women's movement is currently making on the issue of male dominance and negative discrimination against women in the work force. The challenge is being made by women against male dominance in health and psychological care by professional experts in these areas. What is essential here is the consideration that the preeminence of patriarchy and male dominance is being challenged at diverse levels and sectors of our society. We will explore one area of this challenge in depth in Chapter 6, where critical interpretation and gender domination are discussed at the level of psychological interpretation. For the moment, I understand that my discussion of the gender system is all too schematic. My justification is that it is simply suggestive in trying to map out how gender relations reflect the social relations of the personal world in the class structure of monopoly capitalism. It is my contention that a critical interpretation of the personal world must articulate a place for the gender system in its interpretative horizon. To mask out the gender system in an interpretative social science would simply further contribute to the oppression and exploitation already embedded within this system (see Gould, 1976). It would also serve to hamper attempts at a more personal world based on reciprocity and participation by ignoring the oppressive dominance which presently permeates that system and contributes to the present hegemony of a capitalist social structure. The value judgment here is that when the constraints of capital are overriding, there is a vitiation of the personal world which governs the relations between men and women in all social classes.

AGE AND CLASS DYNAMICS

Psychology's major contribution to a discipline dealing with orderly developments with age is predominantly seen in its subdiscipline, developmental psychology (Ausubel & Sullivan, 1970; Ausubel et al., 1980). Age-related phenomena easily lend themselves to biological metaphors, and developmental psychology's history shows strong currents along this line (Ausubel & Sullivan, 1970; Ausubel, et al., 1980). Developmental psychology in the 20th century has waxed and waned in salience, but is clear that since the 1960s it has been considered a discipline at the forefront by psychological practitioners. The Piagetian system of developmental psychology has been a prime mover of this contemporary interest in a concept of development (Sullivan, 1966; 1981).

One of the most pervasive phenomena seen in developmental psychologies of different persuasions is the systematic isolation of developmental phenomena from sociohistorical contexts (Ingleby, 1974). This is done systematically to relieve the conceptual formulation from the ambiguity of all historical phenomena. This leads to a systematic depoliticization of developmental phenomena which masks the ideological nature of the discipline (Ingleby, 1974). In achieving this systematic isolation of developmental phenomena from historical contexts, developmental theories have fallen prey to unconscious ideological problems (Broughton, 1979; Ingleby, 1974; Sullivan, 1977a,b; Wilden, 1975).

Most recently, mainstream developmental psychologists have begun to explore the contextual and political nature of developmental theory (Bronfenbrenner, 1979; Kessen, 1975). There is now a development literature on the psychohistory of children (Aries, 1962; Davis, 1976; De Mause, 1976). There is also an increasing awareness of the influence of economics on the development of psychohistory. The importance of the development of a psychohistory to accord with age-developmental phenomena is that it embeds age phenomena in social history. Age then becomes one factor in historical interpretations and is influenced by the individual and social history that bounds it. Therefore, to talk about childhood is not to raise it to a universal and invariant phenomenon. For example, prior to the Renaissance, children were simply envisioned as miniature adults, childhood *per se* having no special status (Aries, 1962). Childhood became a special status within a particular cultural history. Childhood is also clearly linked to the economic structure of the society (Aries, 1962). Historically, as fathers moved away from domestic settings, childhood as a unique historical period in the life-cycle became more apparent. The relationship between economics and child psychohistory is just starting to be explored with some precision (see Davis, 1976; De Mause, 1974). Since what is being developed here is an interpretative psychology, the development of the discipline of psychohistory is welcomed. This pertains not only to childhood but to all segments of the life-cycle (see Fowler, 1974).

In this final section, we are attempting to relate age-related phenomena to the economic structures involved in class analysis. Not to attempt this linkage would leave these important developmental considerations outside of the purview of a critically interpretative psychology. It is our contention that age-related developmental phenomena must be linked to the larger structural economic totalities. At present, one major social theorist is attempting to link developmental psychological phenomena to the complex class totalities of late capitalism (see Habermas, 1979). In this work, he considers Piaget, Kohlberg, and Loevinger within the context of larger structural totalities perceived to be operant dialectically within individual psychological development. This accents the importance of linking age-related institutions (like childhood, adolescence, and middle age) to the historical structures of the economy as well as such ideological structures as the family and school. A critical interpretation of the personal world would attempt to understand how an age-related institution (for example, childhood) establishes a social legitimacy in the complexity of modern institutions.

In the development of the liberal state, gradualistic social and economic reforms have been focalized on children (Wisby, 1968). Here social reform is achieved by attempting to make the new generation the focal point for societal change. This has not proven to be an effective place for large structural economic changes and, in many instances, acts as a defuser of larger structural changes (De Lone, 1979). The current interest of the state in the child is clearly a way of saying that reform must come gradually and with the next generation. In contemporary life, formal schooling has assumed the role of preparing the child for tomorrow's reforms. This phenomenon has been developing since the turn of the 20th century (Takanishi, 1978). The development of the modern state through the institution of modern formal schooling becomes an important socializer of the class structure (Bourdieu & Passeron, 1977). Schooling and its age-related phenomena act as a conservative force in the maintenance of the social hegemony of the modern state (Bowles & Gintis, 1976; Bourdieu & Passeron, 1977). Social theorists sensitive to the class nature of the modern state are becoming more sensitized to the role of the school in the maintenance of our modern economy and also its reproductive quality (Bowles & Gintis, 1976; Bourdieu & Passeron, 1977). *Reproduction* signifies that the school prepares different children for their places in the class structure and the eventual world of work. This is done by a complex symbolic enterprise which doles out "cultural capital" along the lines of the society's class structure (Bourdieu & Passeron, 1977). Psychology of a critical import must etch out a place for an in-depth treatment of how the personal world is affected by these complex steering mechanisms. Willis's (1977) *Learning to Labour* is one example of a sophisticated treatment of the world of working-class youth and how they are molded by formal schooling for working-class jobs. This study will be discussed more thoroughly in a later chapter as exemplary of a critically interpretative psychology.

CONCLUSION

What has been attempted in this chapter is ambitious, perhaps too ambitious. The reader should understand the importance of mapping out the features discussed herein for a critical psychology with an emancipatory intent. A critical psychology, like all social inquiry with emancipatory impulses, should embrace the task of searching for the new social relations and the new social conflicts that form themselves in a profoundly transformed cultural field (Touraine, 1979). Psychological inquiry as a particular form of cultural action has the task of both denunciation and annunciation (see Freire, 1974). Denunciation is an examination, with critical attention, of those social structures of class, gender, race, and age that systematically exploit and oppress specific cultural forms. Critical psychology pays specific attention to how the personal world of individual and groups is affected by stabilization of unjust social structures. I have attempted to suggest the direction that this might take in the discussions of the tension of habitus and project under conditions of social maintenance (i.e., formative dynamics or the dynamics of the status quo). Some form of class analysis was considered as essential to this endeavor.

The task of critique should not only sensitize one to the deleterious effects on the personal world produced by unjust social structures. Annunciation, in contrast to denunciation, is the critical attempt to explore new institutional projects that seek to overcome structural injustices by attempting to transform social relations toward more equitable states. In many instances, this is first seen in resistance to the status quo. A critical psychology resists interpretation as social deviance, which is the usual direction that mainstream social science takes. In addition to the above, a critical psychology of the personal world attempts to capture dynamics of habitus and project under conditions of real or possible transformation (i.e., transformative dynamics).

INTERPRETATION

> *Believing . . . that man is an animal suspended in webs of significance he himself has spun, I take culture to be those webs, and the analysis of it to be therefore not an experimental science in search of law but an interpretative one in search of meaning.*
>
> —Geertz, 1973, p. 5

THE PERSONAL WORLD AS A TEXT

Ricoeur (1971) has suggested that the model of a text could be most useful in our understanding of human action. In fact, he specifically suggests that meaningful action be considered a text. The understanding of personal human expression that I have developed thus far makes Ricoeur's suggestion quite plausible. Like a text, human action or expression exists to be "interpreted." As I have suggested in Chapter 1, the paradigm of the personal world is communication itself. Consider the possible implications of this in the context of an interpretation of a text. A reader comes to a text to decipher the meaning of the written expressions of an author. Here, the author's expressions are one form of human action: written expression. The text poses a problem to the reader in that she or he must render the author's words meaningful and significant. To consider human actions in the broader sense (written and spoken language, gestures, bodily actions, etc.) as a text to be interpreted means that like a text, human actions are an open work, the meaning of which is in suspense (Ricoeur, 1971). For reasons I have already given in Chapter 1, the shift to a personal metaphor offers certain advantages over mechanical and organic metaphors in the interpretation of human action. However, it is now important to introduce some of the peculiar problems presented to the social inquirer when she or he considers human action as a text to be interpreted. We have already developed the idea of the personal world as a "cultural form" or "form of life." By this, I meant that a cultural form is exemplified in the manner in which a particular social group is connected to the objects, artifacts, institutions, and systematic practices of others that surround it. In other words, the *personal world*, as we experience it, is an identifiable cultural form with some unity and coherence of expression. The question posed now is how an interpreter outside of a particular "cultural form" comes to an understanding of that form of life. This is the "problem of mediation." Some social theorists are skeptical that mediation is even possible (see Bauman, 1978). In essence, they are saying that it is impossible to come to know an alien cultural form. At the other extreme, there is *imperialistic optimism* about the realities other than one's own. This is the position of the "expert–interpreter." Mediation is not necessary here, since all that is demanded is to *assimilate* all alien expressions into one's own system. This form of assimilation presupposes that social realities other than one's own can be bent totally into one's own system of expression. The latter position is arrogant, while the former leads to solipsism. We must avoid these extremes in our attempts to make human actions intelligible. This demands a closer scrutiny of one of the features of human actions discussed briefly in Chapter 2. That feature or mode was referred to as *significance* or intelligibility (see also Schafer, 1976).

Intelligibility

By significance or intelligibility, I mean that human *action* or *conduct* presupposes that the human act could be understood by reference to the ends or intentions an actor is pursuing and also the context or conditions under which action *pursues* its ends (Unger, 1976). The important assumption is that human actions (expressions) are meaningful (i.e., have significance) and are, therefore, not arbitrary. In fact, when an act appears arbitrary, there seems to be a demand in that situation to make it intelligible. Making an act intelligible is, in essence, interpretation. Traditional psychological interpretation has historically attempted to couch the explanation of human action within the ideal of causal explanation. As I indicated in our opening chapter, this type of explanatory device attempts to state how a given set of actions or facts explains another set of actions or facts by linking them together in some temporal sequence (i.e., causal sequence) and attributing one to the other (antecedent–consequent) with more or less probability (cf. Unger, 1976). An interpretative explanation, by contrast, explains by showing that an act makes sense against a background of a social code of rules, practices, or belief. As Unger (1976) puts it, it is a "logic of the situation" rather than a "logic of cause." Unger (1976) insists that an interpretative explanation be distinguished from causal explanation:

> Thus, if we encounter certain features of an artistic style in a painting, we also expect to find other traits of the style in it, though the stylistic attributes cannot be said to either cause or logically imply the other. (p. 255)

Therefore, an interpretative account is an illuminative act rather than an act of causal attribution. It is an illuminative act in that it attempts to uncover the sense of a given action, practice, or constitutive meaning by uncovering the intentions and desires of particular actors (Fay, 1975). It is illuminative because it attempts to uncover the set of rules that make these social practices intelligible and therefore significant. It is an act of analysis and synthesis, as we will see later, in that the act of interpretation attempts to fit these rule-generated and intentional actions into a more total structure (Fay, 1975). By a more *total structure*, I refer to the relational totality of the personal world as a *cultural form*. In interpretative human action within a more total context of cultural forms, the social scientist is redescribing an act or experience by setting it into progressively larger contexts of purpose and intelligibility (Fay, 1975). More is to be said about this, but let me first reflect on the limits of this undertaking.

To suggest interpretative inquiry as a potent method of social investigation does not necessitate a blindness to its limitations. The study of conscious intentions of agents in relation is not to be considered an exhaustive analysis. There are many events in life that escape explanation in terms of our conscious intentions. The substance of tragedy makes this fact a human reality. All those forces that go beyond the reaches of conscious human intentions humble us to the inherent limitation of interpretative methods. Unger (1976) attributes the inherent limitation of this method to a certain fact about human nature:

> The root of the relatively limited range of the interpretative method lies in the dualism of human nature. Man is consciousness capable of intentionality. But he is also a member of the physical world. Though his intentions permeate some aspects of the situation, they never reach all of them.
>
> Whenever we set aside the fact of consciousness, we fall into behaviorism. Whenever we disregard the limitations of consciousness, we slide into idealism. Behaviorism and idealism are the two great sins a method of social study can commit, for both distort crucial traits of human existence. (p. 256)

The vulnerability of the present treatment is to err on the side of *idealism*. This, of course, we wish to avoid. Our bottom-line position requires that the interpreter take the agent's purposes seriously, "to grasp his conduct . . . from the actor's own point of view" (Unger, 1976).

THE PROBLEM OF MEDIATION AND THE PARADOXES OF PSYCHOLOGICAL INTERPRETATION

Is it the task of psychological interpretation, and for that matter any form of social scientific interpretation, to simply understand the world from the point of the agent's conscious intentions and actions? Without precluding a profound sensitivity to the actor's point of view, it seems that the answer to the above question must be no (cf. Giddens, 1979). It seems that simply to reflect the reality of agents involved in their own terms would just be a reiteration and repetition of an already achieved common sense. Left at this level, it would seem that psychological interpretation would be an unnecessary repetition of common sense. There would seem to be no point in this redundancy. Historically, psychology as an interpretative system has eschewed common sense and in, the guise of "expertise," treated it as a low-level reality to transcend. In other words, psychology as a natural science systematically attempted to go beyond common sense. This creates its own problem, and we are left with a riddle:

> If we disregard the meanings an act has for its author and for the other members of the society to which he belongs, we run the risk of losing sight of what is peculiarly social in the conduct we are trying to understand. If, however, we insist on sticking close to the reflective understanding of the agent or his fellows, we are deprived of a standard by which to distinguish insight from illusion. (Unger, 1976, p. 15)

What must be involved in interpretative explanation is a dialectic of distance and relation with the phenomena studied. When the dialectic is collapsed on the side of *distance* (i.e., expert viewpoint), there is the possibility of a total *alienation* from what is studied. At the other extreme, when the dialectic collapses on the side of *relation*, there is such a total immersion (if that is possible) that the interpreting observer has difficulty in separating the forest from the trees. In other words, the task of the psychological interpreter is not simply coding expression from the actor's point of view. If the interpreter is to add anything to the situation beyond the actors themselves, there must be some *recoding* or *resymbolization* of the expressions of the actors (Bleich, 1978). The dialectic of distance-and-relation already alluded to becomes important here because:

> Interpretative explanations requires the interpreter to take the agent's purposes seriously, to grasp his conduct, as has often been said, from the actor's own point of view. But for the observer, the social theorist, or historian to understand a subject's behavior meaningfully, he must be able to decipher what the subject is saying and then recode that message into the language of the observer's own culture. In other words, the greater the distance between the observer and the observed, the more important and the more difficult it becomes to translate from one symbol system into another. (Unger, 1976, p. 257)

What Unger is alluding to is what I will be calling the "problem of mediation," to which we now turn.

Psychological Interpretation as a Double Hermeneutic

All interpretations are mediated through language. We are, therefore, dealing with symbol systems that are public or potentially public. In other words, the reality of communication is that it is, by its very nature, public and verifiable. Here we would like to indicate that interpretation is a symbolic expression which attempts to understand the expression of another person or group. This is a symbolic world and we are locked, as was said at the outset, in the "prison house of language." But is psychological interpretation the same as interpretation at the level of common sense? We have already said no to this question, so we must explore how systematic psychological interpretation differs from interpretation at the level of common sense. Here I would like to introduce a notion coined by Giddens (1979) called the *double hermeneutic*. By this notion, we mean that all systematic interpretation is two-tiered. It expresses the dialectic of *distance-and-relation*, as follows. First of all, systematic interpretation must be related to a particular cultural form and its unique expressions so that action may be appreciated from the point of view of the conscious intentions of the actors involved (i.e., relation). Second, systematic interpretation must be done at enough distance from specific cultural expressions to provide critical feedback on these expressions (i.e., distance). We would call this combined process "resymbolization." If there was no necessity for *resymbolization*, then there would be no need for systematic interpretation of the social, scientific kind. Psychologists would simply be like Job's comforters, reiterating the obvious or making things worse.

The reality of multiple interpretations indicates that what we call the "factual world" is an interpretation. The meaning of human expressions is not an inherent feature of objects or persons, nor is it a psychic event in the head of the actor (Bauman, 1978). The ability to *resymbolize* events or expressions is indicative of the reality that interpretation is a constructive act rather than a process of discovery (Bauman, 1978). Our understanding of a normative social science is simply based on the premise that human actions and intentions are performed within the context of *social rules*. Following Hart (1951), we would contend that interpretative explanation is *ascription* rather than *description*. To talk about intentions or to ascribe intention and purpose is not to talk about private evidence. The "intentional explanation" that ascribes intentions and purposes to human expressions is a language act that is *public*. This type of ascription made on the expression of others will now be discussed as *resymbolization*.

Freud as Resymbolizer. At this point, it is fair to give the reader a concrete example of resymbolization. For psychologists, Freud is an opportune example. Psychoanalysis is a paradigmatic example of an interpretative science. In the psychoanalytic relationship of the patient–therapist, it brings out the notion of the double hermeneutic alluded to above. The analysis is an interpretation of the patient's symptoms that are the symbolic expressions of his or her psychological malaise. The symptoms as presented are opaque, incomprehensible, alien, and so on. The therapist takes these symptoms over time and, in the process of the relationship, attempts to *resymbolize* by offering an alternative framework to the symptoms as originally presented. At the beginning, the distance between the patient and therapist is wide and there is little relation. Over time, the chasm of this distance is bridged (relation). At the same time, this bridging does not lend itself to total identity, whereby the therapist would also have to become ill. The dialectic of distance and relation allows the therapist to identify with the patient's history and, at the same time, resymbolize this history from a safe distance. So we see here the notion of systematic interpretation as involving both identification (relation) and detachment (distance). The therapist's interpretations are a resymbolization of the symptoms in a more complete context of their origins, which heretofore was short-circuited by the patient's repression. The adequacy of the interpretation (resymbolization) is assessed by its utility in the patient's life. Therapy helps a patient create a much more conscious means of self-regulation (Bleich, 1978). Thus, an alternative interpretation has an adaptive communal function. Experience is resymbolized into a form with which consciousness may take more initiative (Bleich, 1978). In our terminology, it makes the patient more of an *agent* and less of a *patient*.

THE HERMENEUTIC OR INTERPRETIVE SPIRAL

In talking about the interpretative spiral, two considerations bear mentioning at the outset. First, psychological interpretation is a *relational* act and, second, it is an act that achieves a dialectic between *analysis* and *synthesis*.

Relational Quality of Interpretation

The interpreter enters into a dynamic relationship with the interpreted. To be systematic, psychological interpretation must be disciplined by a certain distance, but this distance is premised on the assumption that in order to understand something one must relate to it with some degree of affinity or sympathy. Therefore, part of the relationship between the interpreter and the interpreted must be personal (I–thou). The bottom line of this is that the psychological interpreter must seriously take into account the world as it is seen from the agent's sense of his or her intentions. In other words, part of the interpretative moment is an attempt to identify the agent's intentions through a sensitizing to his or her *expressions*. Therefore, part of a systematic interpretative process is a compilation of the first-person account of actors. The psychological interpreter enters into relation with those interpreted by becoming sensitive to their expressed intentions. He or she is not there to manipulate, as in experimental psychology (i.e., I–it relation), but to understand and extend the interpretative world of all concerned.

Interpretation as a Dialectic of Analysis and Synthesis

In the first chapter I talked about the principles of analysis and synthesis. I indicated there that mechanical metaphors were premised on the assumption of analysis (i.e., the whole is the sum of its parts). I also indicated that structuralism and other organic metaphors were premised on a principle of synthesis (i.e., the whole is more than the sum of its parts). Finally, I indicated that a personal metaphor involved a system of interpretation that was a dialectical relation of both analysis and synthesis. I will now illustrate this dialectic by discussing the notion of the interpretative *spiral*.

Hermeneutic Spiral. Hermes was "the interpreter"; hence, the science of interpretation was called *hermeneutics*. The notion of a hermeneutic spiral is well known in the area of philology; we will be using here a more colloquial expression, *the interpretive spiral*. Let us consider first the notion of the interpretative spiral when applied to the interpretation of a text:

> The meaning of a text (or anything else) is a complex of sub-meanings or parts which hang together. (Whenever the parts do not cohere, we confront meaninglessness or chaos, not meaning). Thus the complex of parts is not a merely mechanical colloca-tion, but a relational unity in which the relations of the parts to one another and to the whole constitute an essential aspect of their character as parts. That is, the meaning of a part is determined by its relationship to the whole. Thus, the nature of a partial meaning is dependent on the nature of the whole meaning to which it belongs. From the standpoint of knowledge, therefore, we cannot perceive the meaning of a part until after we have grasped the meaning of the whole, since only then can we understand the function of the part within the whole. No matter how much we may emphasize the quasi-independence of certain parts or the priority of our encounter with parts before any sense of the whole arises, still we cannot understand a part as such until we have a sense of the whole. Dilthey called this apparent paradox the hermeneutic circle and observed that it was not vicious because a genuine dialectic always occurs between our idea of the whole and our perception of the parts that constitute it. Once the dialectic has begun, neither side is totally determined by the other. (Hirsch, 1967, pp. 258–259)

This quote is warranted because of its clarity and precision. The dialectical relation of whole to part and vice versa is equally apposite for human expression in general. As I have indicated in Chapters 1 and 2, human action as expression cannot be considered as an isolated unit. Moving from the analogy of the text above to the interpretation of human action as expression, I think one can interpolate from the text to action quite easily. Thus, one could say the follow-ing: The meaning of human action is a complex of submeanings or parts which hang together. This "hanging together" is the fact that human expression is carried out in a personal world of specific cultural forms. Thus, an act can appear meaningless if the interpreter has no sensitivities to the specific cultural form where the action or expression is operating. Thus, a specific human act is meaningful not in itself but in relation to the larger cultural totality in which it operates. A simple analytic attitude obstructs our understanding of human action as expression. As we have indicated in Chapters 2 and 3, the human act must be considered part of larger structural totality that we have identified as the personal world realized in specific located cultural forms.

Resymbolization and the Problems of Horizons

If psychological interpretation or, for that matter, any social science of interpretation simply reiterated and repeated the life world of some particular cultural form, it would be redundant. One could call such interpreters "scribes" rather than interpreters, since nothing new would be added to the situation. It would seem that there would be no pressing necessity for psychological interpretation. Going native, if you will, produced nothing other than what the natives produce themselves. We have already indicated that the function of resymbolizing human action does more than simply catalogue an expression. I would contend that the resymbolization involved in psychological interpretation is similar to the role of a translator of a text. Gadamer's position is that in text translation:

> The interpreter, like the translator, must capture the sense of his material in and through articulating it in a symbolic framework different from that in which it was originally constituted as meaningful. And as the translator must find a common language that presumes the rights of his mother tongue and at the same time respects the foreigness of his text, so too must the interpreter conceptualize his material in such a way that while its foreigness is presumed, it is nevertheless brought into intelligible relation with his own life world. In Gadamer's terms, a successful interpretation entails a fusion of horizons. (McCarthy, 1978, p. 173)

It is clear that there can be no such thing as the *correct interpretation*, because it will vary according to how the horizons meet one another. These meeting points can be considered in the context of what Lonergan (1972) has called *complementary, genetic,* and *dialectical* horizons. One can say that a horizon of interpretation is complementary when the interpreter and the interpreted are located within the same or similar cultural forms—for example, a middle-class industrial psychological professional studying middle-management executives. It is likely that these horizons will be complementary and, therefore, the resymbolization produced by the interpreter will not be drastically different from or disconfirming of the interpreted (see Ricoeur, 1973).

The case of genetically different horizons presents a different problem for interpretation. Here it is assumed that the interpreter is developmentally superior to those whom she or he studies. Here one can use the example of adults to children, which developmental psychology presents us with as a given reality. In child development theory and research, it is almost universally assumed that the adult has preemptive status; so in an adequate resymbolization of children, therefore, their world is resymbolized (i.e., interpreted) within the context of some adult normative system. Genetic differences in horizons cannot be restricted, however, to developmental psychology. One can see these assumptions operating in cross-cultural research studies of class differences, gender, and so on. Here again, there is an assumption that may or may not be warranted.

Horizons of interpretation may be dialectical. Here one can anticipate clear conflicts in interpretation between the interpreter's *resymbolization* and the interpreted. Here, for example, a psychologist whose cultural and political horizons are not in synchrony with an industrial business outlook may resymbolize a middle-management executive's world in a way that is conflictual to the latter's world view. Here the resymbolization can be initially abrasive rather than complementary.

Given the possibility of multiple horizons between the *interpreter* and the *interpreted*, it would be foolish to venture that there must be one *correct interpretation*. One must deal with conflict of interpretations due to multiple interpretation and consider their differential adequacy. As we shall say later, this question of adequacy must be carried out in the context of discourse and argument. It is not self-evident. Given this *vulnerability*, the whole process of interpretation must be carried out with a considerable degree of humility and openness. This humility and openness is characterized as a dialogue between the interpreter and the interpreted. It is a dynamic relationship, involving a dialectic of "question and answer" (Gadamer, 1975a).

The Dialectic of "Question and Answer"

The notion of psychological interpretation precludes coming to the situation to predict and control what is being observed and interpreted. The interpreter comes to *dialogue*; the assumption being that there are at least two words spoken in an interpretative situation. In a dialogue there is what one might call a "logic of question and answer" (Gadamer, 1975a). When an interpreter comes to a situation, a certain discourse must be started with those being interpreted which involves a revelation. The discourse is intended to reveal something requiring that the whole sequence be opened up by a question (McCarthy, 1978). Conventional psychology, as I understand it, is not drawn to wonder, which the position of a questioner connotes. To come to a situation with a questioning attitude assumes a certain openness on the part of the interpreter. It is clear that the openness of a questioning attitude is not boundless. As Gadamer (1975a) says, it is limited by the horizon of the interpreter. Thus the asking of the question implies both openness and limitation (McCarthy, 1978). As we indicated earlier, the model here is communication itself as seen in a conversation or discourse. An example from an anthropologist illustrates the dialogical dimensions of questions and answer:

> When I went to Morocco in the summer of 1975 my aims were uncertain. I had been to Morocco as an anthropologist twice before, for two years from 1969–1971, and for two months in the summer of 1973. I had attempted, over that time, to fit these experiences to the demands of academic anthropology and I had found that fit unsatisfactory.
> If nothing else, I was determined during the summer of 1975 to confront that dissatisfaction directly. I did not intend to carry out a narrowly defined research

project nor, at the other extreme, would I make the hopeless attempt to "go native." I would simply spend time with people I had come to care about and to enjoy myself with and who seemed to feel similarly towards me, I would try to be sensitive to my needs and theirs, and I would seek to assess my doubts about academic anthropology.

As the summer progressed, the more aware I became that both for myself and for Faqir Mbarek, a 60-year-old Moroccan cultivator whom I had become very close to, our most satisfying project was a series of tape-recorded interviews. These had begun somewhat accidentally. Initially—and I had only one interview in mind—I simply intended to ask him a few general questions about his past and about his attitudes toward his present life. Our second interview was prompted by the unannounced visit from a regional leader of his religious brotherhood. By this time, the form had begun to "take hold": I found myself continuing to single out "events" and Faqir remained more than willing (sometimes suggesting it himself) to sit and talk with me about them. (Dyer, 1979, pp. 216–217)

Recursive Meaning

As with a text, the meaning of human action as expression cannot be established once and for all. This is true because human action expressed in specific cultural forms is dynamic. Horizons, whether from the point of view of the interpreter or the interpreted, are subject to change. Therefore, there are no fixed limits in an interpretation. The limits, at any one time, are fixed by the history of the participants. But history is subject to change by new events. As the anthropological example illustrates, the limits of an interpretative horizon are subject to change when dialogue is present and operating. The interpretative spiral is another way of saying that meaning is recursive. In a text, to say that meaning is recursive means that a text does not establish meaning once and for all; meaning at any one point both depends upon and calls into question what has preceded (Dyer, 1979). The interpretive moment differs markedly from the controlled manipulation to which psychologists are accustomed. The notion of dialogue demands a relational sensitivity as with text interpretation:

> Questions of recursivity can, in principle, be raised at all points in the text. More generally, if we wish to hold to the recursive nature of the fieldwork experience as we pose questions of the text, we must resist irretrievably distorting the text by cutting it into bits and pieces or by playing freely with its actual sequence in time (just as we have tried to resist doing the same to the fieldwork experience by making it into a text). Bearing this in mind, we must ask to what extent this text retains a wholeness, a substantiality, which reflects the recursivity of the experience and yet does not, at the same time, discourage the reader from responding actively to the text. (Dyer, 1979, p. 218)

Reflexive Interpretation

As with text interpretation where the reader assumes an active role in the interpretation of the work, a social scientific interpretation of human action as expression involves the interpreter *actively*. I have already indicated that interpretation is not a passive recording process. In this sense, then, the interpreter is actively involved in a *transaction* with those *interpreted*. The interpreter is not only involved with the actions of others; the interpretation itself must also be considered an act and expression of an agent. Under these conditions the horizon of the interpreter becomes problematic. By problematic, I mean that the horizon of the interpreter must be part of the reflective horizon of the whole interpretative process. Coming to the interpretative situation as an *active agent*, the interpreter must abandon the traditional stance of *expert as distanced rationality*. As an *active agent*, one does not expect mere repetition of the interpretative horizons of the participants. One assumes that a social scientific interpretation of an event will move beyond the point of view of the participants (i.e., resymbolization).

   We will consider the adequacy of an interpretation (resymbolization) shortly. For now, let it be understood that one expects the interpretation to be creative in the sense of challenging the common sense. An interpretation is reflexive when the social scientist ponders and reflects on the effects of his or her horizon on the interpretative events. We are not just agents but agents in relation. It is here that the analogy with the text breaks down. In the text, the interpreter speaks to the text by the interpretation of the work. The author, however, does not speak back. When meaning is *recursive*, as in a conversation, we expect that the resymbolization not only challenges the participants but that the resymbolization can also be challenged by the participants. Therefore, some reflection on one's interpretation seems reasonable under the conditions I have just mentioned. It is, in the classical sense, humility (i.e., knowing where one stands). In coming to the situation of attempting to understand the personal world of others (i.e., their project), one must, therefore, reflect on the effects of *bias*. Different problems are presented when the interpretative horizons differ, that is, when the interpretative horizons of the interpreter appear to be complementary, genetic, or dialectical when related to the project of the participants. When the horizon is complementary, is it because the interpreter has not attempted to achieve some creative distance from those interpreted? When the horizon is dialectical, is it because the interpreter is so *distanced* from the project of the interpreted that the oppositional nature of the interpretation is simply gross misunderstanding? When an interpretative horizon is considered genetic, as in developmental psychology, to what extent is the interpretative system of the adult *adultocentric* rather than the system being a reflection of the interpreter's maturity? These are simply some questions that can present themselves when self-reflection (reflexivity) is operating. This and other dimensions must be considered in the next chapter, where the notion of critical interpretation will be developed.

6

CRITICAL INTERPRETATION

What is critical interpretation? Why does psychology need to go beyond interpretation, which, in short, is a study of the constitutive meaning of actors? This chapter will attempt to deal with these questions.[1]

Let it be said initially that a critical psychology does not reject the hermeneutical. By hermeneutical, I mean the interpretation of the meaning of actions from the actor's point of view. If there is such a thing as psychology, then the interpretation of this microcosm cannot be relinquished. It must be understood that critical interpretation does not relinquish the conscious intentions of actors. In fact, a critical interpretation of the personal world is grounded, at the outset, in the "intentional project" of actors or agents. As I will make clear shortly, this is not the whole matter. The presence of the intentional project as part of a critical interpretative system indicates interest in the concept of human freedom. As I have already indicated in Chapters 3 and 4, the projective character of human action has as one of its constituent features the projection of alternatives to the present and past (i.e., freedom). An interpretative psychology commences to be critical when it attempts to elucidate and criticize those features of a human situation that frustrate intentional agency (i.e., the project). This point is emphasized by Fay (1975) as follows:

> The critical model asserts that in order to have a subject matter at all the social scientist must attempt to understand the intentions and desires of actors he is observing, as well as the rules and constitutive meanings of their social order. The reason for this is that—a critical theory is rooted in the *felt* needs and sufferings of a group of people and, therefore, it is absolutely necessary that a critical theorist come to understand these actors from their own point of view, at least as a first step. (pp. 93–94)

If we stopped here we would be guilty of a crass form of idealism. The full scope of institutional living cannot be reduced to conscious intentions of agents.

---

[1]I wish to bring the reader's attention to a new book by Philip Wexler, *Critical Social Psychology*, which covers the social psychology literature in a more detailed manner. It is published by Routledge & Kegan Paul (1982).

A critical interpretative psychology takes cognizance of the fact that much of human action is outside the conscious control of personal agency and is embedded in social conditions outside human consciousness. These are what I will be calling "structural" as opposed to "intentional" conditions for human action. They were referred to in Chapter 4 as structures of domination. These structural determinants are class dynamics, sexism, racism, and adultocentrism. Basically, these structural dynamics are related to intentional action as habitus is to project. Notwithstanding intentional action, human action must also be understood as being caused by social conditions over which the agent exerts no conscious or intentional control. This limits the scope of the project and, therefore, human freedom, since it is important to understand that social structural conditions unconsciously operate as potent determinants of human action. A specific example will elucidate my point. The work of Brenner (1976) concludes that fluctuations in the economy, specifically in the level of unemployment, are the single most important source of changing rates of admission to mental hospitals. This statistic is also related to class structure. For example, the lower one's economic standing, the greater the cumulative impact of economic dislocation and unemployment. Where economic dislocation is present, psychiatric indices go upward. Mental illness drops when employment goes upward, exclusive of therapeutic intervention. Brenner develops a complex line of arguments, but what is important for our purposes is the finding that structural conditions in the economic sphere that are not directly related to intentional control of the actors involved affect mental health.

This finding raises a note of caution which is part of a critical interpretation of human action. The caution is to be careful in a work that fosters an understanding of intentional human action, like the present work, not to weight agents with more agency than they really have at their disposal. This would allow one to treat a complex condition such as unemployment as something under the intentional control of agents. This type of individualistic psychology ends up by "blaming the victim" for a lack of intentional control. We would call this oppressive rather than critical psychology. What is important in a critical psychology is to make some linkages between the structural dynamics of class, race, sexism, and adultocentricism and the projects of human agents embedded in these historically constituted structures. A critical psychology here will suggest a praxis.

> This means that the quasi-causal explanations which are given must be related to the felt needs and sufferings of the social actors in such a way that they show how these feelings can be overcome by the actors coming to understand themselves in their situation as the product of certain inherent contradictions in their social order, contradictions which they can remove by taking an appropriate course of action to change this social order. A critical social theory is meant to inform and guide the activities of a class of dissatisfied actors—revealing how the irrationalities of social life which are causing the dissatisfaction can be eliminated by taking some specific action which the theory calls for. (Fay, 1977, pp. 97–98)

Thus, the constitutive interest of a critical social psychology will be a core interest in human emancipation and liberation. It is, therefore, a liberation psychology.

EMANCIPATORY PSYCHOLOGY AND ITS RELATION TO A CRITICAL THEORY OF SOCIETY

Emancipatory psychology should not be considered a mere buzz word. As I am defining it, it is a psychology with an expressed interest in the possibilities of human freedom and emancipation. Therefore, one of the features of an adequate critical interpretative psychology is the incorporation of this emancipatory intent. Although this intent may seem strange to the ears of many psychological professionals, it is not without precedent in the social sciences outside of psychology. There are critical theoretical efforts in sociology, anthropology, and political theory that have a significant maturity and presence in these disciplines. Theologians also have developed a theological concern called "liberation theology" (Gutierrez, 1973). To talk, then, of an emancipatory psychology is not to talk in a vacuum but to link the intent of a psychology interested in human freedom with a wider critical social theory. This is one of the goals that must be attempted here.

Psychology and Knowledge-Constitutive Interests

In our opening chapter on psychological metaphors, I maintained that a personal metaphor would have to be interpreted in the context of a critical interest. There I discussed the work of Habermas (1972) which bears review at this point. Habermas schematically represents (see Table 9) social theories as expressing three types of knowledge interests and specifies their ontological orientation and the types of inquiry.

In Chapter 1, I have already indicated how these interests map out in psychological concerns. In a certain sense, the last chapter focussed mostly on

TABLE 9.   Knowledge-Constitutive Interests According to Habermas

Ontological elements of self-formative process	Knowledge-constitutive interest	Type of study or inquiry
1. Labor (instrumental action)	Prediction and control	Empirical–analytical sciences
2. Intention	Understanding	Historical–hermeneutic sciences
3. Authority (power)	Emancipation	Critical theory

the second interest, which involved the meaning and rule structure of actors from their own point of view. Thus, the knowledge-constitutive interest was understanding the meaning of intentional action. Habermas (1970) suggests that hermeneutical sciences are limited in scope because they ignore the role of authority and power in meaning constitution. When power is brought into focus, one moves into an emancipatory interest with critical moments. The critical moment simply means that the power is distributed unequally over different societal groupings, and this has a tendency to frustrate the possibility for intentional projects. Without rejecting a hermeneutic–interpretative orientation, it is essential to understand that the notion of human action is tied to power in a logical manner (Giddens, 1977). Giddens develops this idea by venturing that power represents the capacity of human agents to materialize resources for intentional outcomes. These resources constitute the means to achieve outcomes for intentional action (Giddens, 1977). Giddens then goes on to represent power in two senses. In the first sense, power is seen as the transformative capacity of human agency; that is, the capability of the actors to intervene in a series of events so as to alter the course of a situation. In this first sense, I would say that power is the presence of an intentional project.

This notion was introduced by Von Wright (1971) to draw attention to the possibility that the intentions of agents intervene in larger societal processes to affect the course of events. Intentional intervention is, therefore, the agents' capacity or power to alter events in their world. It is important to distinguish between personal powers and liabilities, since the former assume the agents' capacities for intervention (i.e., freedom), whereas the latter connote a state of patienthood as opposed to agency: ". . . 'having the power' involves being in a certain state and in the case of human individuals, being an agent . . ." (Harré & Secord, 1972, pp. 247–248).

In addition to agent concepts, the description of human life requires a full complement of patient concepts. These we shall call, generically, "liabilities." Unlike powers, liabilities may manifest themselves either because of internal changes or because of external influences. One who exercises a power and is an agent must figure as the source of his or her action in his or her or our account of it, while those who succumb to liabilities may either have only themselves to blame, as we say, or it may be that in accounting for the manifesting of the liability we look for the source in circumstances external to the individual.

The notion of intentional intervention, as Von Wright (1971) understands it, is not a naive voluntarism. Intentions are the projective quality of human action. These intentions are not ventured in a vacuum. The sediment of the past (habitus) makes the possibility of intentional interventions limited in scope. Therefore, freedom is at the cutting edge of apparently dynamic repetitive systems. The past has the power to repeat itself. It can change only by the alteration of human intentions. In this context, I see these interventions operating within the dynamics of the dialectics of habitus and project. In this way, one comes up with a paradox. Freedom and determinism are two sides of a coin. Freedom, the capacity of agents to transform the past, operates within the framework of necessity (determinism). Freedom is, therefore, a struggle with the past for an open future. Therefore,

> the notion of intervention puts an end to the intolerable state of opposition between a mentalistic order of understanding and a physicalistic order of explanation. On the one hand, there is no system without initial state, no initial state without intervention, and no intervention without the exercise of capacity. (Ricoeur, 1978, p. 160)

The "exercise of capacity," as Ricoeur (1978) puts it, is the ability to intervene in one's world as an agent. It is here that one can expect to find dialectical differences in interpretative horizons, depending on the scope that one's horizon gives for the understanding of intentional interventions. Horizons of interpretation which speak to that capacity, as I have done throughout this book, may be said to express an interest in the possibility of an emancipatory praxis.

The second sense is more restricted and expresses societal relationships. This *relational sense* expresses the property of interaction and is defined as the capability to secure outcomes, where the realization of these ends depends on the *agency* of *others*. It is in this latter sense that one speaks of men and women having power over others. Giddens (1977) calls this power the "power of domination." I have already given an example of this power of domination in Chapter 4, in my analysis of the black underclass (see Table 6). This sense of power makes plausible our concern in Chapter 4 for the structure of domination in class relations, sexuality, and relations between adults and children (i.e., adultocentrism). What is important for an emancipatory psychology is a critical interpretative system that analyzes those structural conditions in terms of their enabling or disabling the intentional action of agents (i.e., projects). Simple interpretation of the meaning of action does not encompass this task. Fay (1975) faults interpretive social science on this point on four counts. First, "interpretative inquiry" does not examine "structural conditions." It ignores those conditions which give rise to (enable) or deter (disable) action. Second, interpretative models generally neglect in their explanatory system patterns of unintended consequences, for example, the effects of a system as a whole on individual actors without their knowledge that they are sustaining the system (e.g., class structures, etc.). Third, interpretative models alone provide no way to understand structural conflicts within a society (e.g., racial, class, sexual, etc.). Finally, interpretative models neglect explanation of historical change. They are more interested in presently constituted social forms (i.e., forms of life), which accounts for social order, while ignoring social change and personal and societal transformation. To bridge these shortcomings, one needs a way to examine how larger structural totalities can be linked to human agency dialectically. It is to this I will now turn.

Agency and Structure

Giddens (1979) argues that the notions of action and structure presuppose one another and depend on one another in a dialectical relationship. Implicitly, I have argued the same position in Chapters 3 and 4, in my discussion of the dialectical relationship between habitus and project. It is now necessary to extend the ideas on habitus–project by linking it to the agency–structure dialectic. The relation of agency–structure is a structure in itself, which expresses what Giddens (1979) calls a "duality of structure." Giddens (1979) argues that the "structuration" relates to the basic recursive character of mutual social life, expressing the dependence of structure and agency. He explains this duality of structure as follows:

> By duality of structure, I mean that the structural properties of social system are both the medium and the outcome of practices that constitute the system—the identification of structure with constraint is also rejected structure in both enabling and constraining and it is one of the specific tasks of social theory to study the conditions of social systems that govern the interconnections between the two. (Giddens, 1979, pp. 69–70)

Let us now attempt to examine these interconnections within the context of the habitus–project dialectic (see Table 10). In Chapter 3, Table 2, I identified the notion of habitus with socialization and the notion of project with that of social transformation. At both extremes of this polarity, I identified seven key features that were dialectically related to one another.

To speak of dialectical integration, as I did in Chapters 3 and 4, is to say in Giddens's (1979) terms that this is a "dual structure." One presupposes the other. Let us attach to the above polarities the notions of structure and agency. Let us

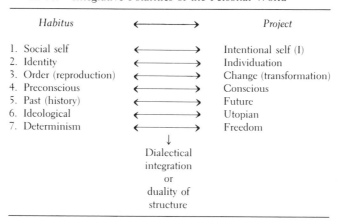

TABLE 10.   Integrative Polarities of the Personal World

	Habitus		*Project*
1.	Social self	←——→	Intentional self (I)
2.	Identity	←——→	Individuation
3.	Order (reproduction)	←——→	Change (transformation)
4.	Preconscious	←——→	Conscious
5.	Past (history)	←——→	Future
6.	Ideological	←——→	Utopian
7.	Determinism	←——→	Freedom

Dialectical
integration
or
duality of
structure

locate structure within the habitus and agency within the project. Yet, let us say that habitus is synonymous with *structure* and *project* is the expression of human agency. These are dual structures because they are mutually dependent on one another for continued life. The problem is that this dual structure and its dialectical relation have been collapsed at one extreme or other in social theorizing, and this has caused a great deal of misunderstanding. Most deterministic theories of human nature collapse their analysis on the habitus–structure side of the polarity. Extreme theories of voluntaristic freedom unencumbered by limits collapse on the project–agency side of the polarity. Following Giddens (1979), I would submit that they must be treated simultaneously and in a more intricate analysis. Let us ponder what is implied here. Our example of infant development in Chapter 3 indicates that the development of a personal world accompanied by a sense of agency is enhanced or debilitated by the structure of the habitus. A disorder induced by chaotic social structures with no stable ideological structures and their accompanying familiarity disables *human agency*. Structures in this sense are ambiguous; that is, they can enable or disable human functioning. An enabling structure enhances human powers and capabilities, whereas a disabling structure detracts from them. Some structures are able to give the agent a sense of personal power that is, in Giddens's (1979) terms, "the capability to make certain accounts count." What a critical social theory must be aware of and take into account is that one person or group's power or project may be another person's or group's bondage. I have already suggested this interrelationship in Chapter 4, in my discussion of class dynamics, gender, ethnicity, and child socialization. Let us now return to these concerns in a discussion of "structures of domination" and "structures of freedom." Let me define a structure of domination as one that hampers human projects and disables the possibility of intentional intervention (i.e., human agency). By contrast, a structure of freedom enhances (i.e., enables) human projects and increases the possibility of intentional intervention. The invariant present in both these structures is power or lack thereof.

Let me now indicate invariant structures and their contribution to the possibilities for human agency. Part of the structure of the habitus is the presence of economic structures. Economic structures are systems of social relations designed for the exploitation of the environment for purposes of human survival. Historically, these structures have varied immensely; the two prominent forms of economic structure today are capitalism and socialism. A second invariant structure is gender. At the basic biological level, this structure of sexuality provides for the replenishment of the human species. The third invariant structure is that of ethnicity. The presence of ethnicity accounts for the universal diversity of the human species. Finally, there is the invariant structure of parenting and adult–child interaction. Because of the sustained dependence of children on adults in the human species, this structure is more characteristic in terms of its impact on the human species than on other animals.

It should be clear by now that these invariant structures are not independent of one another. For example, class relates to ethnicity and they both have implications for child rearing and male–female relations. Although these structures are *invariant*, there are variations in power within these structures that have important implications for the possibility of human agency. An economic structure can be said to be a *structure of domination* when the distribution of power is so skewed to one group that it disables the "projective" character of other groups. This is one of the criticisms of capitalism—that the power is organized so as to enhance or enable one class to the exclusion of others. An economic structure can be said to be a *structure of freedom* when power as enabling is distributed universally and democratically. The same can be said for gender structures. Sexism can be considered a structure of domination within a culture when the projects of one gender (male) are considered more significant than the other (female). Adult–child relations become structures of domination when the dependencies of children are made into invariants (when actually the dependency is variable and will move toward autonomy if encouraged). What is important here is the subtlety of these structures and their interrelation. For example, an economic structure can be considered as liberating (i.e., structure of freedom) at one period and dominating at a later period. The truth of the matter is that we must all live within the ambiguity of social structures. Ambiguity should not lead us into a sterile tolerance which allows us to live with structures of domination because they are intertwined with structures of freedom. The ambiguity of the human condition is that we must all live, as Jesus said, "separating out the cockles and the tares." He said they were intertwined in his parable, but note that he suggests that we must all do the weeding of good from bad. In this sense, an adequate critical psychology must identify and encourage structures of freedom that enhance human agency and correct and help eliminate structures that do not (i.e., structures of domination). Therefore, a critical psychology is in Freire's terms "a joint task of annunciation and denunciation." It is here in the social sphere that value neutrality becomes pernicious. As psychological professionals, we have, it would seem, the moral responsibility to denounce structures of domination that detract from a sense of human agency. Otherwise, we are like Job's comforters, adding insult to injury. This brings us now to a description of how the psychological professional operates within the social order.

PSYCHOLOGY IN THE SOCIAL ORDER

What I am about to discuss is rarely part of the working discussions of psychologists either in theory or practice. It is based on the assumption that psychologists as individuals and institutions are linked to specific "social orders" and that they, therefore, bear the imprint of such an order. Sarason (1981), in the discussion of American social psychology, is quite lucid on this point:

American social psychology has not in any of its major theorizings taken seriously the psychologist's place in the social arrangements and the bearing that place has on what happens to him or her as a person as well as on the social consequences of the psychologist's work. The issue is not in whose domain the problem belongs. It is a central question for every kind of psychologist because every psychologist's thinking and work bear the imprint of his or her place in the social arrangements. And the consequences [of] that work may have to always bear the imprint of those social arrangements. (p. 72)

When one looks at psychology's brief history in the 20th century, it is clear that psychologists have born the imprint of existing social arrangements. In the case of IQ testing in the earlier part of this century, it now can be said, without rancor, that psychologists helped to sustain the racial and social class stereotypes of that era (Kamin, 1974; Sarason, 1981). It can also be said that psychology as a discipline, when following an unreflective course, justifies the existing social order and its power arrangements (Buss, 1975; Zuniga, 1975). Let me elaborate on the last point by a critical analysis of what I will call "the horizon of the expert."

Horizon of the "Expert"

All professionalization assumes as its baseline the existence of expertise. Expert knowledge assumes specialized knowledge and the ability to exercise social distance from the phenomena studied. Specialization also assumes a specific focus and, to be "special" this knowledge must possess uniqueness and identifiable differences from other specialties. American psychology expresses this uniqueness by emphasizing an intimate knowledge of the individual as the unit of analysis. As I have already said in Chapter 3, individualism is a *dominant cultural focus*, and the profession of psychology is a reflection of individualism. The problem of the "expert's" horizon ensues when that "expertise" is not seen as a part of existing social arrangements. This is a difficult task, since the acceptance of this possibility challenges the privileged position of the "expert's position," bringing its traditional position of dispassionate observation and interpretation under suspicion. Be that as it may, there are sound reasons why psychologists are influenced by their position in the social order and why they, in the normal course of events, reflect in their theorizing the existing power arrangements. This is not a static process. The profession of psychology will, in general, reflect the dynamics of a society over time. In a liberal, expansive era like the sixties, the theories did reflect that epoch. In a period of financial retrenchment and conservatism, like the present, one will see theories in psychology that will match the conservatism of the existing social order. These social currents should not be looked at in simple-minded terms. At any one time there are conflicting currents. Psychology as a form of social inquiry will reflect these different currents in its theory and practice. Because psychology seems to be unique among the social sciences in its inability to reflect on its place in the social order, it will, in this unreflective stance, function as an apologist for the status quo.

Because psychology has not traditionally reflected on its place in the social order, it has tended to act within the present cultural horizon as a *complementary horizon* to other horizons that reflect the power of existing social arrangements (i.e., status quo). It has done this indirectly, though in a complementary fashion, in its theoretical and practical interpretations of class structure, gender differences, and racial differences. With psychology's traditional microsocial focus on "individual psyches," psychological interpretations have attributed only to individuals the reasons why there are different power arrangements (i.e., social order). Let me demonstrate this briefly in the context of social class, gender, and race, since they are the categories where I have developed the idea of structures of domination.

Social Class

The position of "expert professional" of whatever color—be it anthropologist, sociologist, or psychologist—involves the production of a specialized language of interpretation of some phenomena and a social distance from the object being interpreted so as to assume an orientation of detached observation. Although detached observation has been assumed in theory, it has never been clearly demonstrated in practice. This can also be said of other disciplines, that is, that it is part of the dynamics of a social order rather than detached from it. Psychology exists in a particular part of a social order and is embedded in one segment of its social strata. A simpler way of saying the above is that psychological professionals are part of the middle class. By education, training, and income, they are embedded in the dynamics of the middle class. One cannot read any simple formula out of the fact that professionals are part of the middle class. What I want to emphasize here is that "professional expertise" will be affected by the dynamics of the class structure in a society. In general, the class horizon of expert–professionals is rarely considered as part of the self-reflexive process of disciplines. In other words, one could say that it is almost in poor taste to think that professionals would be affected by their origins. *Professional distance*, which is a buzz word for *expertise*, is assumed to be operating in all segments of the professional's life.

Historically, psychology as a form of disciplined inquiry is peculiarly tone deaf to its social class origins. It is the one discipline that I know of that has no well-developed self-reflexivity as to its own place in a larger social structure. As a discipline, psychology negates its own class origins and, in practice, ignores its effects in the psychological phenomena that it attempts to give an account of. Sarason (1974), in speaking of psychologists, notes that by self-selection and training they "naturally" think in terms of motivations, attitudes, personality, and many other interpsychic factors. Because, by background and training, their unit of analysis is the individual, their horizon in general masks or conceals the effects of social structure on personal development. Left to their own devices, you might say that psychologists usually attribute social class dynamics to intra-psychic factors (cf. Ryan, 1974). Although rarely reflected upon in this disci-pline, psychology usually follows the movements of the larger sociopolitical and economic context. When the temper is liberal and progressive and the economy is expanding, there is a greater emphasis on transformational (change) models and optimism about the effects of the environment for psychological and social change. Thus intelligence, in Piaget's sense, could be considered as a dynamic transformational activity, and intellectual deficit (cultural deprivation) could be ameliorated by education (i.e., environment; Sullivan, 1981). When the social temper became more conservative, conservative biogenetic metaphors (e.g., Jensen) became more popular.

Gender

Traditional psychology and its many persuasions has always dealt with the question of *gender* as derivative. By derivative, I mean that information about sexuality and gender (usually under the rubric of sex differences) is usually derived from a supposedly larger theoretical construction in which sexuality and gender are accidental by-products of the larger theory. Theory in the above context is an attempt to arrive at some *universal truth* that *transcends* the acci-dental boundaries of *time, space, sexuality*, and so on. Gould (1976) calls this criterion *abstract universality*. She maintains that the traditional philosophical arguments for abstract universality are based on three premises, which she labels essential premises. She frames these premises in the context of philosophy, but I would contend that they are relevant to psychology as a *discipline*, since it is based, as I understand it, on a set of similar premises. Premise 1 states that the appropriate subject matter of philsophy is universal and essential truth that excludes particulars and accidental differences which are, by definition, incom-patible with universal norms (i.e., principles or laws). Premise 2 asserts that in the *human domain*, that is the social, one studies what is common to all human beings and all societies in all times, precluding accidental and local parts as aspects of the human or social (i.e., human nature as such). Premise 3 asserts that *gender* "is an accidental and not a universal property of human beings, since one may be human and not female, or human and not male" (Gould, 1976, p. 8). Gould (1976) draws a specific conclusion from these premises as they relate to women. She maintains that these premises allow one to consider femaleness as an accidental property, thus relegating the question of women philosophicallly as a peripheral concern in the acquisition of *universal abstract* truth. The gender domain is not appropriately universal or essential in the sense required from the above premises.

The premises for abstract universality can be considered as part of the interpretative horizon for much of what we would be naming as conventional social science. This horizon, as I understand it, is dialectically opposed to possible horizons of specificty, where such aspects as *gender* and *ethnicity* are considered accidental or nonessential aspects of truth. Gould (1976) maintains that when one is operating out of a horizon of abstract universality, there are certain masked value premises which must be made problematic by critical scrutiny. Critique is necessary, Gould says, because she feels that the interpetative horizon of abstract universality reflects the interests, needs, and prejudices of particular social groups. There are two reasons why this becomes problematic. First, "such essential prejudices may introduce distortions into our very understanding of what is essential" (Gould, 1976, p. 17). Second, "essentialism reflects such interests in a particularly harmful way, because it tends to mask them under the guise of universality and therefore is deceptive" (Gould, 1976, p. 17). There is no claim being made that the criterion of abstract universality is inherently vicious and conscious in its distortions. What Gould (1976) is suggesting, however, is that in actual practice the criterion of universality seems to choose properties as *essential* and *universal* which reflect the interests of a particular gender (i.e., male) and which were associated with roles and functions in which males predominated (Gould, 1976).

Although she is speaking about philosophical discourse and goes on to give specific instances of male dominance in some of the classical philosophers, one can, nevertheless, draw an analogy for the traditional discipline of psychology and cite specific examples of distortions of sexuality that have been masked by the assumptions of abstract essentialism. In the social sciences, and particularly psychology, the issue of gender is usually discussed under the rubric of what we have called in Chapter 1 "organic metaphors." In contemporary thinking since the beginning of this century, gender has been couched within the framework of biological-evolutionary theory (Shields, 1975). Shields (1975) contends that the "interpretative horizon" of evolutionary theory as it came to be applied to the social sciences was male-dominant in that it essentially assumed the biological supremacy of the Caucasian male (Wine *et al.*, 1980). This interpretative horizon was, therefore, skewed from its beginnings to interpret *one gender* and *race* as the apex of evolutionary development.

These turn-of-the-century functionalist–evolutionary perspectives attributed differential functional differences to males and females, always considering males superior. Earlier discussions of differences in brain size between male and female saw men as superior in intellectual capacity because of their larger brain size. Furthermore, a hypothesis of greater male variability, and its relation to social and educational issues, showed males more flexible. Here the Darwinian assumption establishing that variation within a species is a requisite for evolutionary adaptation and progress was buttressed with the belief that men were more variable and, therefore, superior in their potential development (Shields, 1975). Basically, what we are seeing here in a historical perspective is the development of a universalistic norm that just happens to be male-dominant as well as race-dominant. From what one could call a horizon of "feminist psychology," there is the critical understanding that much of our present discussions are similarly afflicted by views of male superiority (Favreau, 1977). Frequently, male and female differences that are considered as descriptive and nonessential are given universality by attributing them to biological origins. In other words, they are considered as universal because of their assumed biological origins. As I have proposed from the outset, a psychology that is a critical interpretation of the personal world will deal with gender as a historically constituted category rather than assuming the preeminence of biological determinants as found in organic metaphors. This does not rule out the presence of biological determinants in both gender similarity and difference. A critical interpretative horizon will, however, resist attributing historically constituted sex differences as if they were biologically given (i.e., attribute sex difference to the same type of organic biological metaphor). There is a universalistic ring to biological metaphors that interpret sex differences as by-products of some "biological universals where males by *nature* rather than *nurture* are superior to females.

In short, then, what must be suggested in our present treatment is an interpretative horizon that deals with gender as historically constituted where sex differences are considered within the context of cultural dominance rather than the by-product of a genetic heritage. To treat sex differences within a cultural horizon that is constituted historically does not solve problems but simply provides a critical lens which can countenance sexual dominance as an offshoot of a male-dominated culture. This involves an alternative notion of *concrete universality* suggested by Gould (1976), which is a counterposition to the position suggested as *abstract universality*. Sexual difference are not mere accidents that are assimilated into biological universals. Gender here is put in the category of a concrete universal. The notion of universality is retained, but it exists in radical transformation:

> Whereas the criterion of abstract universality is concerned only with what is essentially the same and excludes accidental differences, the criterion of concrete universality is concerned also with human and social differences, and includes them not simply as accidents but as aspects which constitute the universal or the essence itself. (Gould, 1976, p. 26)

The universal, however, is not permanently couched in biological or organic categories but, rather, within *historically* constituted categories, which I would maintain as consistent with what I have been calling the personal metaphor. It is a psychology from the point of view of agents who constitute their worlds by actions in historical time rather than biological time. Therefore:

> Concrete universality thus implies an alternative conception of the nature of history itself. As against the view that historical events are exemplifications of fixed, a priori essences, this view regards essences, as well as history itself, as constituted or created by the actions of individuals. Concrete universality concerns the present as the moment in which individuals create history through their interaction with each other and with the objective world. It sees the past as the set of interactions which provides the circumstances for present action. This action transforms these circumstances in accordance with the future envisioned by these agents. (Gould, 1976, p. 27)

Thus, from this interpretative horizon, one can speak of cultural projects that are gender-related and dialectical in their relation to one another, as male and female roles have been historically. Women, for example, could be expressing a concrete universality (e.g., "women's truth"). Women become a unique category and a "cultural form," as we have indicated, by having a specific role in history unto themselves but also in dialectical tension with cultural roles of a category called men.

What I am about to say is speculative and controversial, and I would beg the reader to treat it as a plausible hypothesis. From a feminist's perspective, one could argue that the psychological horizons, as we understand them, can be considered as arising historically in western culture under the dominance of white males who have taken themselves to be speaking in a universal human voice (Scheman, 1981). As we have already indicated, the tradition is one of abstract universality, so that the issue of gender, as we discuss it here, is masked from the potential awareness of the position of the universalizer. Moreover, it is contended that this specific type of universalization is the reflection of a liberal ideology (Sullivan, 1977a, 1977b, 1981). *Liberal ideology* refers to a style of thought developed at the time of the French Revolution (Mannheim, 1953a). I ventured to suggest earlier that liberal ideology is the philosophical underpinning of the notion of *individualism* and critiqued this idea in Chapter 3. As we said there, the notion of individualism proceeds with the assumption that larger social totalities are the product of the organization of individual units (i.e., principle of analysis). Thus, it is assumed to be a *natural fact* that individuals exist in essence as separate units, so that it can be said that psychological characteristics such as wants, preferences, needs, abilities, and so on are inherent and located "in individuals" (Scheman, 1981). We have already critiqued this from the perspective defined as the personal metaphor (see Chapters 1 and 3).

I will now add that the horizon of individualism as embedded in liberal social theory carries, in addition, the baggage of male dominance and bias. Male dominance means simply that one would expect interpretative horizons to be markedly influenced by points of view that are related to the preemptive status of males historically in our culture and specifically in the profession of psychology. This assumes that women have quite consistently been excluded from the centrally important metaphysical, epistemological, moral, and political conceptions of personhood (Scheman, 1981). The case is made by feminists that men have traditionally held the power to name or define what it is to be human, adult, intelligent, moral, and so on and have done so in response to their own experience of and need for separateness and distinctiveness (Scheman, 1981). By contrast, traditional female roles have emphasized fusion (Bakan, 1966). If this present work simply emphasized the human agent with intentions and meanings that are self-contained, one could say that the bias of individualism was operating. *Individual agency* as a self-contained structure is not, however, the "horizon" under which the perspective being developed here is operating. Personal agency is agency in relation to others, as I have pointed out in Chapter 3. A psychological horizon that is a critical interpretation of the personal world combines feminist critique of male dominance (i.e., self-contained agency) with the relational totality called community (i.e., persons in relation). In this light, I would hope that the perspective anticipates transformed, equitable gender relationships in the wider culture and more specifically at the level of theory and practice of the discipline of psychology.

Ethnicity

Historically, psychology has not used the construct of ethnicity or race in any direct manner at the level of theory. In fact, most of the classical texts on personality do not have a salient place for this construct. Like gender, it is a particularistic construct that contradicts the direction of theorizing attempting to formulate perspectives which are abstract and universal. By and large, the work of treating ethnicity or race has traditionally fallen to sociologists and anthropologists who treat race within the context of culture.

This in no way says that the profession of psychology has not been involved with the phenomena of ethnicity, for the fact of the matter is that, at the level of social policy, its influence in North American psychology has been prepotent. Historically in North America, it was the psychological testing movement that first brought the profession of psychology in contact with *ethnics*. This was through the screening of immigrants through the use of standardized tests of *intelligence*. In general, it was assumed that the standardized test was a way of setting up a universalistic norm for the assessment of intellectual abilities. The assumption was that these standardized tests were "culture-free," that is, that they were not biased in any systematic way against any particular segment of the population. Thus we see here the development of what we have previously called a *universal abstract* norm. At the turn of this century, these tests were considered by the immigration department of the United States as *objective* norms by which to assess the intelligence of new immigrants (Kamin, 1974). By today's standards, the discussion of ethnic groups by some of the major figures of the psychometric testing movement would be considered racist and inflamatory (Kamin, 1974). Yet, this language was considered objective for its time. In addition, the discussion of the determining factors for the presence or absence of intellectual capacity was embedded in the now shopworn (but far from dead) heredity–environment controversy. That controversy was over the relative weight that one should give these assessed psychological traits in relation to genetic factors (heredity) or the environment. In different epochs, psychometricians have come down on one side or the other on this controversy. In general, either side of this interpretive horizon (i.e., genetic or environmental) has been directed pejoratively toward certain cultural and ethnic groups. In fact, the heredity–environment controversy seems to follow economic currents operating in the social order and the values that are precipitated within an expanding and contracting economy. When the economy is *expanding,* there is an optimism that *environmental* factors such as education can improve the lot of disenfranchised minorities. In the 1960s, for example, blacks were considered *culturally deprived,* and this deficit, it was felt, could be alleviated by education and other services. Within the context of what is a universalistic means, it is interesting to note that blacks were perceived to be lacking in culture (i.e., culturally deprived). When the economy went into retrenchment, one finds that environmental theories were eclipsed, since they held out the possibility that environmental factors such as *health* and *education* benefit disenfranchised groups. When this happens, the more conservative genetic theories come to the fore, as witnessed in the Jensen controversy. Here differences in cultures are perceived to be influenced in a major way by genetic dispositions. This justifies a cutback in services, since they supposedly would not be effective in any event (Hunt & Sullivan, 1974).

The Need for Reflexive Interpretation

What we have just tried to document is the conventional bias of psychology within the social order. When this bias is not reflected upon in some critical manner (i.e., nonreflexive), the psychological interpreter has historically ventured interpretations that reflect his or her class origins and the place psychology has in relation to the culture at large. The position of "expert," by definition, leaves the expert as unproblematic. Expertise problematizes all without itself being problematized. Two assumptions tend to mask awareness of problems that expert positions pose. The assumptions are that the expert has distance from extraneous relationships surrounding a phenomenon being studied and that the expert is in a nonnormative or value-free position (i.e., unbiased). As we have hinted in relation to class, gender, and ethnicity, the position of nonbias is difficult to sustain when a historical perspective is taken. The bias of psychology,

when left unchecked (i.e., unreflexive), will, in all probability, reflect in substantial ways the biases of the wider culture. In other words, psychological
interpretation will normally be a complementary horizon to the status quo,
psychological expertise functioning as a form of legitimation or rationalization
for the normative order or status quo. Does this mean then that bias of interpretation makes the task of psychological interpretation impossible? We think not. We
think that the interpreter must reflect a bias of viewpoint. We would project that
a critical psychology would critically reflect in its bias of viewpoint, problematiz-
ing itself in relation to the phenomenon studied (i.e., reflexive psychology). The
mapping of structure, as I have defined it, onto agency allows us to examine
power relations in society and the conflictual nature of power in society. Although psychology as a professional discipline has a commitment to even-tem-
pered observation and interpretation, it cannot consider itself a neutral umpire in
the power dynamics of the society in which it is located. Silence or expressed
value neutrality gives consent.

I have just indicated how the value neutrality simply serves the needs of the
social order and stabilizes the power dynamics within a society. An "emancipatory psychology" with critical intent will serve and help *name the world* (Freire) of
oppressed groups and, in essence, serve the cause of freedom. In that sense, order
without freedom (domination) will be *criticized* rather than *apologized*. One
would, therefore, expect that there would be conflict of interpretations between
psychological professionals, depending on their identification with the social
order where they are located. For example, *resistance* is almost always considered
by social scientists as deviance when it comes from a group that has little social
power. If you consider my extended example of the black underclass in Chapter
4, one finds that the author of this book criticizes conventional social science
interpretations of resistance because they are always labeled pejoratively (i.e.,
deviance) when they challenge a dominant economic system. A critical psychology must examine resistance in a more subtle manner. Resistance can, and
many times is, the beginning of an authentic project which is indicative of a
developing sense of agency. Resistance becomes an agent process when it fosters
the development of community and solidarity. When a personal world appears to
be sustained by resistance, one can sense that there is an authentic cultural
project, which is not deviance but a community of what one might say is a
different order. A critical psychology will judge resistance by its fruits rather than
its inconvenience to a particular social order (i.e., deviance). When resistance
fosters authentic community, that is, the presence of a personal world (I–thou
relations) and in the process challenges an order of domination, it is to be
celebrated by a critical psychology. In general, with psychology's focus on the
individual (microsocial) while ignoring conditions in a wider social order (macrosocial), the structural background conditions for actions are left unanalyzed. I
will demonstrate this by example in the next chapter. The fetish of a microsocial

focus puts the burden of virtue and vice totally on the *individual* while ignoring the larger social conditions in which human actions are embedded. At the same time, a critical psychology with emancipatory intent must look at resistance critically. Resistance is not a virtue in itself. It must reach a level of community consensus to seriously challenge the dominance of an unjust social order. A focused rebellion and resistance simply encourages counterviolence by the larger social order and, in essence, justifies its continued presence. There must be attractive features in resistance groups which make them real alternatives to some of the destructive features of the present order. When this occurs, one can say that agency transforms the structural conditions rather than being molded by it. Resistance, and the corresponding projects that are suggested by it, must be able to challenge unjust structural conditions to be considered an effective praxis.

In another vein, one can look at the structure–agency interplay in relationship to economic conditions and mental illness (e.g., Brenner, 1973). I have already discussed Brenner's work in several places because it establishes a clearcut relationship between economic conditions and mental problems. What happens when a major economic activity leaves an area suddenly, as is the case with what is now being called "the runaway factory"? Capital, when left unchecked, moves where labor costs are cheapest. Under these constraints, corporate activity (when capital needs are premptive over community needs) tends to leave areas where labor presents problems for profits. What is becoming clear today is that communities with high employment have much lower rates of mental illness when compared to marginal economic communities (Brenner, 1976). This suggests to me a relationship between favorable structural economic conditions and the development of a personal world, where a sense of agency pervades (i.e., agent–structure interplay). The presence of community projects is related to lower mental illness problems in that community. A community economic activity must then be seen as an *enabling structure*, which enhances the agency of community life. What has now become apparent by the "runaway factory" is the whole question of economic (corporate) responsibility. Can the state simply let the corporation serve the needs of capital and have it ignore the needs of the community? When this operates, the state sends in psychologists, psychiatrists, social workers, and so on to pick up the pieces after the corporation has gone on to greener capital pastures. If psychological professionals simply look at the microsocial world of the individual, they end up by treating individuals for mental illness, deviance, and other forms of disabling maneuvers. As I have indicated several times already, this type of attempt ends up "blaming the victim." A critical orientation would take a different tack. Instead of treating individual symptoms and labeling them, it might involve itself in community mobilization to make corporate interests sensitive to the needs of the community. Instead of treating reactive depression, alcoholism, and so on and ameliorating symptoms in a structural vacuum, critical psychological intervention will make the linkages between disabling social structure and personal psychological malaise. Psychology's microsocial focus here advocates the personal world of the client over the impersonality of structural decisions. As MacMurray says, the personal world must not be the servant of impersonal structures (MacMurray, 1961). Impersonal structure must serve the personal world and enable viable community.

CAUSE AND INTENTION

The concrete exemplification of an interpretative system which honors the distinction and dialectical relationship of agency and structure raises a final issue of the relationship of cause and intention. I have identified cause with determinism and intention with freedom. The dialectical relationship indicates my understanding that freedom is a struggle with necessity. I take this struggle as a given for the human species where the ordered necessity of institutions is created by the free appropriation of a community consensus. Thus, for humans, the deterministic qualities of institutional structures of order are at their incipient stages, created by a form of free consensus. To talk about human causes is to clearly distinguish them from the mechanical causality of Hume (Von Wright, 1971). To talk about the structural determinism of human institutions which are historically constituted, it is suggested that the term *quasi causal* be used (Von Wright, 1971). Von Wright (1971), in his seminal work *Explanation and Understanding*, ventures that when talking about quasicausal factors one is indulging in *explanation*, whereas when one is talking about intentions for action the concept of *understanding* is more apt. In the above sense, explanation is oriented toward *structural* explanations, which attempt to achieve a description of a phenomenon in terms of the structures such as class, ethnicity, gender, and so on. Understanding, by contrast, is the interpretative grasp of a lived situation which is supported by *intentions* (Von Wright, 1971). If the duality of the agent–structure dynamics holds, then *understanding* and *explanation* must be dialectically related to one another in a critically interpretative psychology. The proof of the pudding is in the eating. I will close this book with several examples of research that exemplifies this process to a greater or lesser extent. The works are only suggestive and reflect my biases and background. It is with these examples that we will deal with the notion of adequacy of an interpretation.

7

ADEQUACY OF INTERPRETATION

To give an account or interpretation of a situation from the point of view of a systematic social scientific interpretation is to place human action as expression (i.e., the project) in an understandable context (Winter, 1966). As in the logic of text interpretation, the meaning of human action must be understood as *a limited field of possible interpretations or accounts* (Ricoeur 1971). One thing certain can be said when considering accounts: that there are no finalized accounts. Recursive meaning, as I have explained it, means that the demand for final, absolute, unrevisable truth cannot be satisfied. As a text never has one interpretation, so also human action is open to multiple interpretations and is partially contingent on the viewpoint of the interpreter. Conflict of interpretations is the first systematic problem of an interpretative psychology. If one faces the problems of "conflict of interpretations," it gets us into the thorny problem of the "bias of the interpreter," which conventional psychology attempts to ignore. Instead of ignoring bias and treating it as an abberant process in scientific inquiry, an interpretative psychology takes bias as part of its problematic. We are calling bias the "interpretative horizon." The use of the term *bias* in this context is not pejorative. The notion of bias simply is the procedural assumption that the interpreters bring with them a certain horizon of expectations—that is, of beliefs, practices, concepts and norms—that belongs to their own life world. I have already looked at the bias of interpretative horizon in the previous chapter, when I discussed the conventional location of psychology.

We must now consider why the problem of horizon or perspective raises serious problems for most contemporary psychological systems of interpretation. Most contemporary systems act consciously or unconsciously within an absolute viewpoint. A viewpoint is absolute when it is assumed that the limits of bias are not operating. The absolute viewpoint demands that relativity be overcome for the sake of stabilized substantial knowledge. In its most dramatic form, relativism opens a Pandora's box for any perspective that assumes, without question, the priority of its own horizon of interpretation. When this box is opened, it raises the possibility that all limited viewpoints have limited access to an event and are, therefore, equally suspect. This is a severe problem for an interpretative psychology, since it can lead to a total agnosticism or subjectivism. We must leave open the possibility that some interpretations are better than others under certain conditions and in certain situations. As I shall point out later, *deciding on the validity of interpretations must be something that must be argued convincingly by an interpreter.*

The *limits* of the observer's perspective are what I am calling the "horizon." *Horizon* calls forth the notion of *perspective.* "Perspectivism" does not renounce the postulate of objectivity and the possibility of arriving at decisions when there is conflict in interpretations (Lonergan, 1972; Mannheim, 1953b). Objectivity within the perspective of perspectivism is an indirect means for dealing with conflict of interpretations. Therefore, to accept the relational quality of knowledge does not mean that *objectivity* is abandoned. *Relationism* simply asserts that every interpretative statement can be relationally formulated, that is, formulated within the horizon of the interpreter. Relationism need not collapse into relativism:

> It becomes relativism only when it is linked with the older static ideal of eternal, unperspectivistic truths independent of the subjective experience of the observer, and when it is judged by this alien ideal of absolute truth. (Mannheim, 1953b, p. 300)

THE PROBLEM OF MULTIPLE HORIZONS

Multiple horizons open the door to conflicts of interpretation. Conflicts invite resolutions. To resolve a conflict *objectively* is to involve oneself in an adjudication process. But differences in interpretations do not always invite conflict. It is only under certain conditions that horizons of interpretation are in conflict, and it is important to distinguish when these conditions are present or absent.

Lonergan (1972) maintains that horizons of interpretation can be complementary, genetic, or dialectical. First, complementary horizons are different horizons that complement rather than conflict with one another. For example, theories formulated between mechanical metaphors of interpretation, discussed in Chapter 1, are probably complementary even though each separate discipline brings its own separate and idiosyncratic concerns. They are complementary in that they share a similar and complementary horizon on what constitutes evidence, what is adequate theory, and so on. In the sense of a complementary social science perspective, one could also say that the structuralism of Chomsky's linguistic theories, Piaget's genetic epistemology, and Kohlberg's moral stage structure share complementary epistemological horizons.

Second, horizons or perspectives can also differ *genetically.* Here we are talking about how horizons change over time for a given interpreter. A dramatic example is the differences in interpretation one finds in reading Doestoevsky in early college and then again at age 60. Probable differences in interpretation result from different horizons that a person operates out of in two distinct segments of the life cycle. To say that horizons differ genetically means that the differences are related as successive processes of development (Lonergan, 1972). The later stages presuppose earlier stages, partly to include them or possibly transform them. Piaget's stages of intellectual development or Kohlberg's stages of moral reasoning illustrate developmental genetic horizons, which they call stages. Each stage within a developmental context can be considered as an interpretative horizon. The horizons are said to be related *genetically* insofar as all higher stages integrate and transform the horizons of early stages. Since these stages are earlier and later, no two are simultaneous (Lonergan, 1972). Unlike complementary horizons, which are part of a single communal world, genetic horizons are to be seen more in the context of a single biography as a single history (Lonergan, 1972).

Third, horizons can differ *dialectically*. Here, horizons can conflict with one another to a point where they are in *dialectical relation*. This means that the presence of one horizon proposes the negation of another, and vice versa. Here we must face a reality in social inquiry: namely, that the social sciences house interpretative systems which are radically in conflict with one another. To say that a difference is *dialectical* is to understand that differences in some interpretative horizons are not simply linguistic in nature. I have hinted in the first chapter that interpretative horizons based on mechanical metaphors differ radically from those that propose metaphors of the personal. At the level of both theory and research, they differ in a radical sense in epistemological and normative world view. By accepting that interpretative horizons may differ dialectically, one accepts the problematic presented in conflicts of interpretations. One does not totally abandon a desire for truth in this situation; what has to be abandoned is the desire to be in position of the "absolute truth." If truth is relative, it is not necessarily arbitrary. We not only face the problem that interpretations differ over time, since they also differ between interpreters. The latter situation simply illustrates the selectivity of all perspectives when facing one another. Winter (1968) maintains that "the selectivity of perspective does not mean that one view is wrong and another correct; it means only that varying perspectives on a problem illumine different aspects" (p. 168).

It is understood that a social scientific account is different in perspective from the point of view of those interpreted. Given that it differs, one of the first characteristics of an adequate account is its negotiability. This constitutes the first condition for an adequate account.

1. *An adequate account is negotiated.* As Bauman (1978) points out, the epistemology of an interpretative horizon cannot be detached from the sociology

of communication. Here a "horizon of expert" (i.e., nonnegotiated interpretation) is challenged. It is not a foregone conclusion that a social scientific resymbolization is an adequate interpretation; the dictionary says that to negotiate is to
communicate or confer with another so as to arrive at a settlement of some
matter. Since a social scientific account of human action (i.e., the project) is
concerned with delineating the conditions and interests that entered into the
constituting of the project (Winter, 1966), it would simply appear courteous, at
least, to clear the account with those whose project has been interpreted. In other
words, the interpreted must be able to identify themselves in the account given.
If this identification is not possible, it does not necessarily mean that the interpretation is inadequate. It may be that the interpretation reveals something to
the participants that they do not want themselves or others to see. This may not
be conscious. It may be, however, that the interpretation is inadequate; that it
simply is a misunderstanding of the nature of the project interpreted. It could be
a mixture of both. The negotiated account at least gives the reader the knowledge
that the account being given may be problematic and in need of revision. In the
interest of truth, it gives the reader a chance to be part of the judgmental process.

2. *An adequate account presents itself as an argument.* One must accept, at
the outset, the inevitability of conflicts of interpretation. The nature of the
conflicts can vary from situation to situation. One interpretation may be rejected
by the participants. If accepted by the participants, it may be questioned by other
scientific perspectives. For example, throughout this book I have argued that an
adequate account of the personal world as a cultural form cannot be enhanced by
biological metaphors. The fact that biological metaphors have a compelling
attraction for many social scientists means that they cannot be categorically
rejected. Alternatives must be argued and reasoned to be considered plausible.
Nothing, and that includes what is clearest in one's own argument, can be
considered a foregone conclusion. As we have just indicated, all conclusions are
negotiable. All interpretations must labor under the possible handicap of circularity, that is what is proved in the conclusion was assumed in the beginning
(Hirsch, 1967). Therefore, interpretative accounts of necessity must entertain
alternatives. To entertain does not necessarily mean that alternative arguments
must be embraced. Arguments for an interpretation proceed under the assumption that alternative interpretations may be related to one's own in either a
complementary, genetic, or dialectical manner. One argues for an account on
the basis of its validity. Here we are using the term *validity* in a restricted sense.
A valid argument does not connote verification. To speak of verification is to
assume that an account's conclusion is true (Hirsch, 1967). *Validity* connotes a
more humble stance in that one tries to show, in arguing an account, that the
interpretation is plausible. But validation goes beyond mere plausibility. In accepting conflicts of interpretation, one does not have to accept that accounts are
of equal merit. Validation demands adjudication of conflicts:

> The job of validation is to evaluate the disparate constructions which understanding
> has brought forward. Validation is therefore the fundamental task of interpretation as a
> discipline, since whenever agreement already exists there is little practical need for
> validation. (Hirsch, 1967, p. 170)

It is here that it is appropriate to make a distinction, infrequently made, between *understanding* and *explanation*. Ricoeur (1978) contends that understanding and explanation are relative moments in the complex process called interpretation. When interpretative horizons are complementary or identical, there is very little need for explanation. Explanation is warranted only when there is a situation of conflict. Thus, explanation follows because:

> Understanding calls for explanation as soon as there is no longer dialogue, where the
> give and take of questions and answers permit an interpretation to be verified as it
> unfolds. When I do not spontaneously understand, I ask for an explanation and the
> explanation that you offer allows me to understand you better. (Ricoeur, 1978, p. 153)

Here I would agree with both Hirsch (1967) and Ricoeur (1978) that validation is a test of our guesses at interpretation, following a logic of probability rather than a logic of empirical verification. In a certain sense, an interpretation is an argument for, or advocacy of, a point of view or horizon. The idea of advocacy is unknown to psychologists and would probably be considered in a pejorative sense if left unexplained. My contention is that a *valid* argument is a form of advocacy for a particular interpretation. In advocating a particular interpretation, one brings forth factual evidence in favor of a particular interpretation and unfavorable to one's opponents. Validation, in these terms, can be seen as an argumentative discipline comparable to the juridical procedures of legal interpretation (Ricoeur, 1978). Arguments for an interpretation must be seen in a relative light. This does not in itself lead to skepticism. One expects that interpretations can conflict with one another. In arguing for a particular interpretation, one develops evidence to show that an interpretation is *probable* under these circumstances. When an interpretation is in conflict with another interpretation of human action, the better interpretation attempts to show that it is more probable than its adversary under the circumstances. Part of that argumentation is the practical significance of the interpretation. The practical significance of a critical interpretation is in its specification of a praxis of freedom.

3. *An adequate account expresses an emancipatory praxis.* Our concern over the practical significance must not be identified with "instrumental rationality." I have already expressed my concerns about theories whose practical significance is to manipulate and control others. Our concern is that a critical interpretation have the practical significance of enhancing human freedom. This is basically what I mean by an emancipatory praxis. What is most characteristic about a personal metaphor is its emphasis on the historically constituted nature of the social world. A good social scientific interpretation reveals the nature of the

social world as constituted by *agents* in *relation*. Our assumption has been throughout that we are not organisms or mechanisms but *persons*. The personal world is a world constructed in *history*. An emancipatory account will reveal to the participants the fact that their world is constructed *historically*. To show that a world is built through a history is also to indicate that it can be changed through history. The personal world and its specific expression in diverse cultural forms is, therefore, seen in the light of a constituted history. I have indicated this in Chapter 3 where the dialectis of *habitus* and *project* were considered. This dialectic is important here because it prevents us from giving a facile account of human freedom and human determinism. In considering the dialectic of habitus and project, one does not consider freedom and determinism as operating independently. I would contend that human freedom by free agents happens in an *ordered* (habitus) context, order providing the footing for historical projects. Therefore, one condition for freedom is an ordered environment. Freedom (i.e., projects) is built on the *habitual order* of cultural patterns. Therefore, an arbitrary social order—that is, a cultural situation without habitual patterns (order)— is an enemy to human freedom. This is anarchy in the pejorative sense and the situation invites dictators. Order, when one forgets that it is the historical constitution of human agents (i.e., entrenched patterns), becomes the enemy of human freedom. Here order is seen as some kind of mechanism or the achievement of a biological process (i.e., mechanical and organic metaphors).

This detracts from the human constitution of cultural formation which is the real historical construction of a personal world. To interpret *order* in this manner is to limit the scope of *human projects*. Cultural forms are not just made and repeated (i.e., order), they are also changed and transformed by agents who dream new dreams (i.e., projects). To deny the openness of historical projects is, therefore, a form of *oppression*. I use the term *oppression* here to connote the possibility that some interpretations mask or conceal the scope of human possibilities. An interpretative account, with emancipatory intent, will attempt to express the dialectic of order and change of a specific cultural form. Its practical utility is that it will attempt to portray the resilient nature of habitual patterns, that is, show how history has formed a group through their own agency and continues (i.e., determinism). At the same time, it attempts to express the projective (toward the future) nature of those forms, which is an opening on the future (i.e., transformation). Therefore, the emancipatory account is couched in a language that expresses the possibility for change. The minimum conditions for an account of this nature are that the account show (1) how the immediate social and physical conditions that determine and structure the social world (i.e., habitus) are *mediated* through the agency of actors who have some control over the constitution of that world (i.e., responsibility) and (2) how they take for granted understandings that are also mediated and, therefore, could be otherwise (Corrigan & Willis, 1980). This type of an account is an emancipatory praxis, which, in the words of Marx, not only interprets the world but also suggests how to change it.

4. An adequate account is critical. I have already indicated that social scientific accounts are *resymbolizations.* Thus, one does not expect reiteration of the point of view of those interpreted. Resymoblization brings something new to a situation. It characterizes the expressions of others in a new light. As long as accounts are considered negotiable, as in criterion 1, there is no reason for a systematic account of others to be in complete agreement with those interpreted. Consider the example of psychoanalytic interpretation. The patient comes to the analyst and spends a considerable amount of time expressing the understanding he or she has of existing personal problems. In many instances, the symptoms presented appear overdetermined; that is, they repeat themselves insistently and the patient appears to have no control over them. Part of the analytic treatment is to *resymbolize* the patient's symptoms (i.e., analyst's interpretation) by reinterpreting those symptoms and pointing out their historical origins. The analyst's interpretation frequently meets with resistance and denial by the patient. An analytic interpretation is considered useful if it expands the scope of the patient's horizon, and the eventual acceptance of an analytic interpretation is accompanied by a decrease in overdetermined symptoms. At a microsocial level, one might label a critical interpretation part of the emancipatory praxis of criterion 3.

At the very least, a critical account of the personal world of human actors must accomplish two objectives. First, in resymbolizing the world of others, it should sensitively reflect back to them and others their capacities and powers for *intentional intervention.* This expands the scope of freedom in a positive direction by naming those powers. Second, it also draws attention to the liabilities or impediments to human agency. These liabilities can be psychological factors that are part of a person's makeup and history. They can also be external, that is, persons or institutions that block the agent's intentional interventions. It is here that the bias of horizon (e.g., class, gender, ethnicity) may have deleterious effects. For example, the class nature of a psychological interpretation may find a certain rebellion of lower-class children simply representing personal liabilities. In a more sympathetic light, one could consider rebellion as a resistance (intentional intervention) to arbitrary and unjust social systems. Or the bias produced by gender domination (i.e., male dominance) in psychology may consistently cast women in the role of patient and men as agents. Let us end this chapter with two specific examples of what I would call interpretative studies and conclude by judging them on the four criteria I have just described. I have chosen these two studies in order to help the reader understand how I employ the adequacy criteria. The studies are also suggestive because they lend themselves to comparisons and illustrate the notion of conflict of interpretations. Both studies interpret the actions of working-class boys who could well have been part of a sample of working-class youth. The fact that they are so similar as a population or sample makes the differences in interpretation grounds for asking questions about adequacy in comparative terms.

Study 1: *The Rules of Disorder*

Background

As the authors (Marsh *et al.*, 1978) indicate at the outset, the ostensible subject matter of this book is violence and disorder. Carried out in Great Britain, this is essentially a study of lower-socioeconomic-class males. The settings for this interpretative enquiry are the classroom and the "terraces of football grounds." The authors contend that the systematic interpretative inquiry they offer indicates that the boys in these settings demonstrate none of the anarchy and impulsiveness attributed to them in the popular press. Instead, the interpretation they offer indicates a consistent rule-orientation that belies an attribution of anarchy and impulse. Marsh *et al.* (1978) indicate different interpretative horizons between media interpretations and the interpretations the boys ascribe to their actions. The media interpretation of these youths casts them within a rhetoric of impulse and anarchy, their behavior toward one another appearing to be senseless and opaque. Marsh *et al.* do not propose to repeat either of these rhetorics in their interpretative scheme. Their aim, as they see it, is to discover the content and organization of the social knowledge they believe an individual member has to have in order to be able to perform reasonably well in the social life of his or her group. They make it clear at the outset that they are not proposing a general performance theory based on knowledge of rule structures (i.e., competence). They maintain that in order to accomplish a *performance model* they would have to introduce the *concept of intention* at the level of action. This is important to note here, because what is absent from the outset is the possibility of assessing the possibilities for intentional intervention into the action sequence.

These authors also make a distinction between rules of interpretation and rules of action. Their position is that in order to assert that conduct is orderly, one must demonstrate that it is "rule-governed." Rule governance does not have to be modulated by conscious intentions. In fact, rules are held (often tacitly) by all members of a group or community as representations of legitimacy and acceptability. The rule is a *normative* concept; implicit in the notion is the directive of *ought* and *should*. These authors maintain that rules are not statistical abstractions but are directly discoverable in the accounts that football fans and school troublemakers offer in explaining what they do and why they do it. The strongest test for the existence of a rule is the ability of a member of the group to articulate it. These authors venture that an articulation constitutes a "reason for action." In addition to an articulation, a rule can also be considered present when *breaches* in a rule are articulated (e.g., this is wrong, etc.). The rule is seen in consensus. Consensus connotes social order. Social order presupposed that members of a group have a common knowledge of the rules (Marsh *et al.*, 1978).

Method

Intensive Design. An intensive design is one in which a few members are selected to exemplify a class. This is to be distinguished by a sample (i.e., extensive design) where a sufficiently large number of cases are collected to be statistically representative of a class. Marsh *et al.* (1978) claim that their option for an intensive design is hazardous, but they feel the eventual payoff is worth the risk:

> In the ethogenic approach we follow the example of chemists and anatomists and adopt the intensive design, that is, we undertake a detailed study of a few cases selected as typical. We are sure of the detailed validity of our analyses of our cases. We can only hypothesize that they are typical. Those who would follow the extensive deisgn can be sure their results are typical, but they must hypothesize as to whether the few properties that survive the working of the inverse ratio between extension and intention have an individuality at all. Our football following participants are from one group of fans, supporters of just one local club. With few exceptions, the participants in our study of talk about classroom volume and its accounting are from one school that is graced by some of the fans. We are inclined to think that they are pretty typical, given our less detached work at other football grounds and what we know of other groups of school children. (Marsh *et al.*, 1978, pp. 20–21)

First-Person Accounts. In an earlier work (Harrè and Secord, 1972) one of the authors (Harrè) had a chapter entitled "Why not ask them." There, as well as in this work, a rationale for the importance of first-person accounts is given. The authors feel that in interpreting human action, the best (although not ultimate) authorities as to the action are the actors themselves. Therefore, a sensitivity as to how the actors view their actions is encouraged. They acknowledge the real possibility for personal self-deceit and blindness but nevertheless feel in the long run that a "match emerges gradually between what is seen to happen and what is said to happen" (p. 21). Therefore, they state:

> We take it axiomatic that unless it can be established to the contrary, the best authorities as to what went on are the actors themselves. Their meaning and their rules have priority in the scientific analysis of the phenomena. (p. 22)

A methodological ideal is to combine with the first-person accounts of human action an access to the situation in which the accounts take place (e.g., schools and terraces). One is combining here interview data with some limited form of participant observation.

Concepts Utilized. Four concepts are utilized in this study. They are violence, meaning, aggression, and order. *Violence* connotes an action of physical interference with another, whether or not it is being mediated by the use of a weapon. Examples go from bullying and biting (nonmediated) to a display of a weapon (mediated). Further violence is symbolic when there is a mimicry of interference (e.g., threats). To talk about a notion called "symbolic violence" brings them to a second concept, labeled *meaning*, which is the interpretation of action (either movements or speech) as intelligible because of its relation to the intentions of actors. Meaning, as they interpret it, is achieved by setting up a relational structure between meaningful element of the symbolic system. Thus symbols can stand as the real thing for action in a symbolic system (e.g., clenched fist). Most symbols of violence, if carried out, would lead to death or serious injury. The symbol of violence short-circuits this ultimate outcome and is, therefore, seen as a ritualized aggression that prevents serious injury. The third concept is that of *aggression*. Aggression is distinguished from violence by attributing a biological origin to aggression. The authors maintain that there is a biogenetic origin to aggression. Aggression is, therefore, a biological universal. Aggression as a biological universal takes various historically conditioned forms. Here, a distinction is made between real aggression from another concept called "aggro." Aggro is a ritual manifestation of aggression in symbolic form. The importance of this distinction will be seen later when these authors take a final excursion into sociobiology. Finally, *order* is the concept that highlights the assumption that human action follows rules and is governed by consensus. These authors contend that a test for the presence of order is the degree to which the unexpected does not happen and that expectations are fulfilled.

It should be noted that there is no extensive discussion for the selection of these concepts as part of their interpretation format. I will say more about this later in criticism.

The Schoolroom Setting. There is no elaborate description of this setting. The authors maintain that they are concerned to explore the interpretation and genesis of disorder and violence in the classroom from the point of view of the pupil. The main shortcoming (the authors acknowledge it) of this part of the investigation is that the researchers had no access to the classroom. They are, therefore, somewhat at a disadvantage in validating the veracity of the pupil's account. The authors indicate a lower-class school, and this is corroborated by student descriptions. The students in the class being investigated are non-academic by their own descriptions. They perceive the school as a waste of time and not to be taken seriously. They perceive themselves as "written off," since teachers do not place a high value on them.

"Messing About." "Messing about" connotes a consistent classroom interaction between pupils and teacher. The interaction takes the form of an offence–retribution cycle. The basic sequence is as follows: (1) teacher loses control, (2) students made power play, (3) teacher retaliates to restore order. As exemplified in an interview:

I was in this lesson and two or three of my mates were messing about and I was sat
across the other side of the classroom and someone chucked a piece of chalk at the
teacher and then started shouting his mouth off. Straight away the teacher "knew"
who it was (unjust accusation). He turned round and kicked me out of the classroom
for doing it. So I hit him. (p. 34)

The authors contend that retaliation, like hitting the teacher, is not blind
violence but a maneuver to save a "moral career." The students are very sensitive
to offenses to what they consider their dignity and standing (i.e., moral career).
Moral career is, therefore, a public reputation that is valued by the participants.
A teacher's devaluation is an affront to a moral career. Students maintain that
devaluation occurs in three ways. First, there are teachers who are arrogant and
distant: "There's a few teachers who you can get on with and talk to. They seem
to understand you. But most of the others—it's a nine to four job" (p. 36).

Second, there is anonymity produced by the teachers' not knowing the
students' names. Finally, there are offenses of unfairness where a student is
compared to other students negatively or a punishment unrelated to the offense is
given.

Marsh *et al.* (1978) draw parallels between the order cycle of the school and
that which takes place in the home. The students find very little consistent
valuing in school or at home. They develop their "moral careers" on the terraces:

You've got to decide whether it's a freer atmosphere and rowdyism or silence and total
obedience, because I don't think you have any feelings towards somebody in schools
where you just have to sit in silence in sort of neat rows. (pp. 52–53)

The Football Grounds (Terraces)

Although many strata of society will watch a football match, this study
focuses only on lower-socioeconomic-class youth in this setting. The authors
maintain that in this setting there is an order known to the police and the youth
which demands that specific locations be adhered to by home- and opposing-
team fans. The police place themselves as a wedge between the home-team fans'
territory and that of the opposition team.

Method. Besides interviewing, Marsh *et al.* made video recordings at the
setting. They maintain that the recordings reveal patterns and groupings in
which they were able to attribute a number of salient behavioral and social
characteristics to the boys.

The boys looked at in this study ranged from 12 to 17 years of age and a
specific grouping wore "aggro outfits," which will be explained shortly. Mainstream vocabulary would put them in the category of "rowdy." The authors feel
that this is an excellent setting for the development of what they have called
"moral careers." In this setting, a moral career connotes a way of "becoming

somebody" and is a highly structured affair. These rowdy events are seen by the authors as intelligible and rule-following rather than anarchic. The moral career in this setting is an available structure in a youth culture for the establishment of a stable self-identity. It seems to enable these boys to achieve kinds of *reputations* and *images* denied them in mainstream society. In other words, it is an attempt to maintain a certain dignity. The data were gathered through videotapes and also biographical materials taken from samples of fans in each group, with special attention to what they have labeled "rowdies" and "town boys."

Rowdies. The authors identify six types of careers: (1) *Chanters* or *chant leaders* who lead in chanting and taunting. (2) *Aggro leaders*, who lead in the expression of aggression that is not seriously injurious. This is a high-status role and the person must be courageous, but not foolishly so. (3) *Nutters*, who number five or six boys. They are individuals whose behavior is considered so outrageous as to fall completely outside the range of rational actions (i.e., goes mad, goes wild). Marsh *et al.* (1978) see the nutter as one who demonstrates to others what they should not do, providing living proof by negative example. (4) *Hooligans* are boys whose actions are generally thought to be worthy of praise. This is contrary to the media's definition of *hooligan*. Usually, their acts cause small damage to property and enrage some members of mainstream culture. Nutters and hooligans have this in common—they provide entertainment. A hooligan is "a jester." (5) An *organizer* is an older boy who does the management work (e.g., gets bases, etc.). (6) A *scapegoat* is an individual who receives negative attributions (i.e., he is someone habitually picked on).

Town Boys. Town boys are boys who have graduated the rowdy group. They are less noisy but clearly tough. They are also seen as heavy drinkers.

The authors suggest that these career structures help to explain social behavior that on the surface appears irrational. In looking at these moral careers, these authors follow a procedure that attempts to discern the rule structure within this subculture at the terraces. They maintain that you can build a picture of what the rules for action might be by extensive interviewing. The terms of reference involve identifying a stable pattern of ascription of meaning to objects and events.

In the 50 interviews done on these boys in groups of three and four, one point seemed to be recurring, and that is, as the authors maintain, that fights, either in the large group or face to face, are not random. Aggression ensues in circumstances where fans are able to *specify* legitimacy for their actions. To summarize the question of rule structure, it may be said that: (1) There is a set of interpretative rules (related to territorial rules and fouls, etc.) indicating when aggression on rival fans is appropriate. (2) Once a fight has commenced, there is a rule structure governing its course. It should be noted that the rules are protective against serious injury, so that no great physical harm normally comes to the participants. (3) Rules indicating the closing off and termination of fights can be isolated. There is a further subtlety within these structures in that rules can be seen as implicit or explicit. These fans, for example, are not conscious of the full range of significance of the rule structure they are following. The authors sometimes purposely alter the accounts of the participants so as to create a discrepancy that may indicate explicitly the presence of an implicit rule. As the authors indicate: "By looking at what they say is wrong with the story we can obtain recognition of the rule-breaches. This, in turn, allows us to specify what the rules themselves are" (p. 111).

Aggro as Ritualized Aggression. The authors contend that up to this point they have simply stood inside the phenomena and have carefully listened to what the fans have to say. I will later challenge this contention. Their claim is that they have attempted a close scrutiny of the events on the terrace and have attempted to outline what they consider to be the social order of shared meaning that exists there. For the participants, their order as outlined is a structural delineation of rule structure. They speculate at the end of this structural interpretation its limitations:

> The existence of order, then, does not, in itself, guarantee Utopia. The only way of deciding the merit or value to society of a particular order is through reference to the function of that order. (p. 117)

In presenting the notion of aggro the authors make an excursion into functional explanation. The shift here, to my mind, is dramatic. They venture that the *rule structure* they interpret is seen as a reason for action. Their excursion into aggro theory assumes that a functional theory can arrive at causal mechanisms. They feel that questions concerning causes might be legitimately posed at a different level of analysis. The level of discourse with aggro is one that is sociobiological in nature. Their functional explanation, therefore, is couched in an organic metaphor. Here, the authors move from a conventional order based on rule structures that are generated in the interpretation of participant accounts to what the authors call a "natural order." In other words, the naturalistic explanation is clearly an account from the outside, an interpretative maneuver that indicates the preoccupations of the researchers rather than the participants.

Marsh *et al.* see that they are now moving into a primary stage of what they call "explanation." The adequacy of this move should be judged in two ways: (1) On the basis of its fit with the commonsense reality of the participants—in other words, it has to "go down" with the fans when it is fed back to them so that they can see themselves in this type of description. (2) On the terms of criteria that are used in the initial interpretation (i.e., moral careers).

The notion of moral careers is the initial framework that also offers prescriptive rules for violence, but the authors believe it is necessary to go beyond this initial framework to the functional sociobiological framework of ritualized aggression (i.e., aggro). Why? Because they feel it necessary to discuss the function that aggro plays in society at large. This theoretical jump is justified because it places football violence in a larger perspective by showing that their aggression is part of a more generalized mode of ritualized aggression that is species generated. Let us look at aggression as a form of ritual (i.e., aggro). The authors define a ritual in terms of four key elements: (1) a pattern of routines of behavior, (2) a system of signs that convey covert messages, (3) the existence of sanctions expressing strong moral approval or disapproval, and (4) a conventional relation between the actions in which the ritual is performed and the social act achieved by its successful completion.

Marsh *et al.* contend that life in the classroom and on the terraces can be functionally catalogued as a form of ritualized aggression (i.e., aggro). They summarize their work as follows:

> In this book we have not sought to excuse the football fan or the classroom trouble-maker. Instead we have simply tried to show that the events which outrage us have a different reality and are capable of being construed in a very different manner. We have tried to reveal social order in events which are traditionally seen as dangerously anarchic. And social order, whether it be in the form of ritual or not, is something that needs to be recognized and seen as having utility and merit. When magistrates and police refer to fans as animals and savages and when teachers are unwittingly engaged in the processes of systematic humiliation and depersonalization of schoolkids, order is threatened. We may never, given our existing social frames of reference, be able to create a system of schooling which kids regard as relevant to their own culture and socialization. And without doubt, we will be able to suppress entirely the aggression and the striving to subdue rivals that has been characteristic of young males in all human societies at all times in history. Given this, we must look to ways of managing hostility and violence rather than naively hoping that they will go away. If we accept that there are, from one significant standpoint at least, *rules* of disorder, we might be able to develop management strategies which have far more purpose and effect than those which have currently emerged from the atmosphere of moral outrage and collective hysteria. (p. 134)

STUDY 2: LEARNING TO LABOUR

Willis's (1977) work is discussed here for several reasons. It deals with a group of working-class boys in England who are roughly comparable to the population of boys in the Marsh *et al.* study just quoted. The treatment of these boys and the interpretation of their world is markedly different and contrasts rather dramatically with the previous study. In my view, this work represents a critical interpretation of the personal world of white working-class youth. It is a critical study because, to my mind, it self-consciously deals with the duality between *agency* and *structure*. For that reason, it provides an excellent example of a critical account. It also brings home to the reader, by contrast, the real possibility of what I have called "conflict of interpretations." Its similarity to *The*

Rules of Disorder lies in its dealing with a comparable population; also, both studies attempt interpretive accounts from the point of view of the actors involved. The studies depart drastically when they move from first-person accounts to causal–structural interpretations. In *The Rules of Disorder*, as we have already seen, the authors move from ethnography to a sociobiological functional metaphor (aggro) in explaining the action of participants. Willis's work also moves beyond interpretation from the point of view of actors and their agency to a *structural–historical* interpretation, which I would deem to be what I have called the personal metaphor. Where structure is biology in *The Rules of Disorder*, it is, by contrast, historically material in *Learning to Labour*. With that in mind, let us now venture a detailed description of this study and then a critical analysis of both these studies as to their adequacy as accounts.

Aim of the Study

The study was designed to explore the transition from school to work of nonacademic working-class boys. The author poses this question in the form of a partial statement: "The difficult thing to explain about how middle class kids get middle class jobs is why others let them. The difficult thing to explain about how working class kids get working class jobs is why they let themselves" (p. 1).

Willis concentrates his efforts on the latter question by focusing on white working-class boys in an industrial town (Hammerstwon) in Great Britain. Concurrently, he does a partial analysis of the role of gender and ethnicity with this type of specified population. Like Marsh *et al.* (1978), Willis (1977) notes that the existence of this culture has been picked up conventionally and pegged by the mass media in sensational modes of violence and, in schools, depicted as lacking in discipline (i.e., deviant).

Method

The study used multiple techniques of data collection. Case-study work, interviewing, group discussions, and participant observation were used over a period of 2 years as the boys proceeded through their last 2 years at school and also into the early months of work. The author justifies his use of qualitative methods for the following reasons:

> The qualitative method and participant observation used in the research, and the ethnographic format of the presentation were dictated by the nature of my interest in the cultural. The techniques are suited to record this level and have a sensibility to meanings and values as well as an ability to represent and interpret symbolic articulations, practices and forms of cultural production. In particular, the ethnographic account without always knowing how, can allow a degree of the activity, creativity and human agency within the object of study to come through into the analysis and the reader's experience. This is vital to my purposes where I view the cultural not simply

as a set of transferred internal structures (as in the usual notions of socialization) nor as
the passive result of the action of dominant ideology downwards (as in certain kinds of
Marxism) but at least, in part, as the product of collective human praxis. (p. 4)

Let me note parenthetically that Willis's notion of culture shares my assumption that the personal world of culture is both habitus and project. The
research carried out involved one main and five comparative studies. The main
study was concentrated on 12 nonacademic working-class lads, as Willis calls
them, attending a working-class school. This was an all-male school, which was
twinned with an all-girls school in the same area. The lads are depicted throughout as "resistant or in rebellion" against the school system. The author also did
some comparative work looking at lower-class conformist youth ("ear-oles") and
a middle-class population at a residential school. The later groups (i.e., ear-oles
and middle class) were selected to give a comparative dimension to the study
along the parameters of class, ability, school regime, and orientation to school.
The most extensive and intensive work, however, was done with the main group
(the lads), who were observed in the school (participant observation), both in
class and in the halls, during their leisure activities. In addition, there were
regularly recorded group discussions, informal interviews, and diaries. Long
taped conservations were also accomplished with parents, teachers, and school
officials.

Ethnography: The Study of Culture

Ethnography is the study of the peculiar or typical character of a specific
culture. Thus, part of this study is an ethnography of a specific population of
white working-class youth. The ethnographic study constitutes one moment in a
larger, more total analysis. In my terms, *ethnography*, as Willis uses it, is the
study of the personal agency of specific cultural forms, attempting to discern this
from first-person accounts. Three chapters are devoted to this enterprise (Chapters 2, 3, and 4), the final chapters being devoted to structural analysis of what I
and others have called the agency–structure dialectic. In Willis's terms, it is an
analysis of the interrelations between *cultural* and *class-structure* analysis.

Elements of an Oppositional Culture. The *cultural project* of the lads is
their opposition or resistance to the larger middle-class norms and values. Some
of the salient characteristics of the lads' culture, which Willis discusses from
observation and their first-person accounts, are their opposition to authority and
rejection of conformity. The lads see themselves in opposition to larger school
norms and authority (i.e., countercultural). As they say of the conformists (ear-
oles), "It is not so much that they support teachers, rather they support the idea
of teachers" (pp. 13, 14). If the school is the "zone of the formal," with its clear
structure, school rules, pedagogic practices, and so on, the lads' counterculture

is the "zone of the informal." Oppositional culture seeks to go beyond the reach of "the rule." Informal opposition to the school is manifested in the struggle to gain symbolic and physical space and to *defeat* the school's main express purpose to make you *work*.

WILLIS: Do you think you've learnt anything at school, has it changed or moulded your values?

JULY: I don't think the school does fucking anything to you. . . . It never has had much effect on anybody. I don't think (after) you've learnt the basics. I mean school, it's fucking four hours a day. But it ain't the teachers who mould you, it's the fucking kids you meet. You're only with teachers thirty percent of the time in school, the other fucking two-thirds are just talking, fucking pickin' an argument, messing about. (p. 26)

Apropos of work avoidance:

WILLIS: When was the last time you did some writing?

FUZZ: Oh, er, last time was in careers 'cos I writ yes on a piece of paper, that broke me heart.

WILLIS: Why did it break your heart?

FUZZ: I mean to write 'cos I was going to try and go through the term without writing anything. 'Cos since we've cum back, I aint dun nothing (it was half way through term). (p. 27)

Thus, the countercultural infrastructure of the informal group makes possible a distinctive type of class contact for the lads; Willis contends that it is a class culture which is distinct from the dominant one. The main objective of the counterculture is to defeat formality by informal modes of interaction. When the culture becomes fully developed, as in the later years of schooling, it becomes quite adept at managing the formal system and limiting its demands to an absolute minimum.

Part of this resistance is done with humor: "We can make them laff—they can make us laff." Even where there is a violent tone to some forms of interaction, it is seen as an *orderly* process, concurring with the position on violence taken in *The Rules of Disorder*. At the same time, the orderliness of violence is given a different interpretation by Willis:

> It should be noted that despite its destructiveness, anti-social nature and apparent irrationality, violence is not completely random, or in any sense the absolute overthrow of social order. Even when directed at outside groups (and thereby, of course, helping to define an in-group) one of the important aspects of violence is precisely its social meaning within "the lads" own culture. It marks the last move in, and final validation of the informal status system. (p. 35)

Note the different construction of the nature of violence interpreted here with the sociobiological notion of aggro in the previous chapter.

Finally, the group defines itself through what Willis identifies as sexism and racism. Two other groups are set up in opposition and against whom the lads define themselves by opposition by maintaining a clear sense of superiority in relation to girls (women) and ethnic minority groups. Women in this context are seen as sexual objects and comforters. The whore–virgin categorization is marked with these boys. Girl friends or eventual spouses are set in a different category from the "easy lay." A girl friend "must be attractive but not experienced." The lads are very much prone to racial slurring. In Great Britain, such slurs are levied against non-Caucasians—Asians and West Indians.

Racism allows these boys the luxury of looking down on groups more marginal than their own.

Class and Institutional Form of Culture

Willis attempts to make linkages between the counter-school culture and the shop-floor culture. Therefore, the setting for some of his fieldwork is at the factory, where the lads will eventually work, and in the home, to do interviews with the parents. It is here that the author contextualizes the boys as a specific *form of life* that relates school to shop floor to home. A form of life or culture must be seen as partially self-constituted. Although Willis acknowledges the alienating and restrictive conditions of work, he does not reduce the working-class form to simple passive puppeteering:

> They exercise their abilities and seek enjoyment in activity even when most controlled by others. Paradoxically, they thread through the dead experience of work a living culture which is far from a simple reflex of defeat. This is the same fundamental taking hold of an alienating situation that one finds in counter-school culture and its attempt to weave a tapestry of interest and dimension through the dry institutional text. These cultures are not simply layers of padding between human beings and unpleasantness. They are appropriations in their own right, exercises of skill, notions, activities, applied toward particular ends. (p. 52)

Thus, common cultural themes or institutional class forms, as Willis calls them, emerge between counter-school culture and shop-floor culture. (1) Chauvinism, which is masculine toughness exemplified by the lads, is reflected in one of the central themes of the shop floor (male chauvinism). (2) Information control, a main theme of the shop-floor culture, is mirrored in the counter-school culture, where the lads attempt to take control of classes, substitute their own unofficial timetables, and control their own routines. (3) Informal grouping, characterized by the counter-school culture, is also seen as a fundamental organizational unit at the shop. Willis maintains that, at the level of informal groups, there is a zone where there are strategies for wresting control of symbolic and real space from official authority. (4) Idiosyncratic language and humor is highly developed both on the shop floor and in the school. Along with verbal humor, there is a physical humor—the practical joke. At the factory, as in the counter-school culture, many jokes circle around the concept of authority itself. Finally, (5) practice is more important than theory. The rejection of schoolwork in counter-school culture is matched on the shop floor with the massive feeling characteristic of the working class that practice is more important than theory.

Willis is at pains to articulate these themes as part of a salient form of working-class life. The type of resistance embedded in these themes departs drastically from what might be a middle-class theme. As Willis points out:

> When the middle class child is thrown back on his indigenous culture, instead of finding strengthening and confirming oppositional themes there, he finds the same ones. Centripetal forces act to throw him back to the institution. (p. 76)

Willis (1977) maintains that the lads, as a located culture, systematically prepare themselves for a certain kind of work, and this marks them out from the ear-oles (i.e., conformists not only in schoolwork but also in occupation). They (the lads) scoff at the vocational counseling provided by the school. There is a direct link between school and work in the minds of the lads and their parents. Particular job choice does not matter too much to the lads:

SPANKSY: I think that we . . . more or less, we're the ones that do the hard grafting, but not them, they'll be the office workers. . . . I aint got no ambitions, I don't wanna have. . . . I just want to have a nice wage that'd just see me through. (p. 98)

By contrast, conformist working-class youth (ear-oles) have a much greater identification with teachers' authority—career choice, and so on. The ear-oles probably buy into white-collar semiskilled jobs. A resistance to mental work becomes a resistance to white-collar authority for the lads. Here the split between white- and blue-collar worker begins. The split is carried out with some agency, and this must clearly be understood because it is consistent with a choice ideology that is espoused in liberal democracies. So Willis notes:

> It is only in the uncovering of this subjective assent that we will understand their behaviour in a way which properly presents their own full powers, and appreciates the contradictory, half real notion of freedom at stake for them (and others) in a liberal social democracy. (p. 103)

SPIKE: . . . it gets me mad to see these kids working in a fucking office. I just dunno how they do it, honestly. I've got freedom, I've got . . . I can get money, it's hard to explain. (p. 104)

In giving these examples, Willis concludes that the counter-school culture of the lads smooths the way to manual work.

Willis ends his ethnographic section with a paradox and, in my terms, posing the dialectic of agency and structure:

> It is they, not formal schooling, which carry "the lads" over into a certain application to the productive process. In a sense, therefore, there is an element of self-domination in the acceptance of subordinate roles in western capitalism. However, this damnation is experienced paradoxically as a form of true learning, appropriation, and as a kind of resistance. How are we to understand this? (p. 113)

Going beyond Ethnography: Analysis

Part 2 of *Learning to Labour* is entitled "Analysis." Willis claims that it is necessary to go beyond classroom and workplace ethnography to a more structural analysis. The interpretative tools that Willis will utilize draw from a Marxist class analysis. Thus, analysis, for Willis, is a class-structural analysis. Combining class analysis with ethnography makes this study unique and classifies it as a Marxist ethnography. It is here that Willis attempts to embed individual agency (culture) with the larger structural class dynamics of capitalism (structure). In that sense, it honors what Giddens has called "the duality of agency and structure." The interplay between agency and structure is introduced with two structural analytic concepts, *penetration* and *limitation*. These concepts become important when Willis summarizes and projects problems that issue from the ethnography:

> Although we have looked in some detail through case study at the experience and cultural processes of being male, white, working class, unqualified, disaffected and moving into manual work in contemporary capitalism, there are still some mysteries to be explained. . . . We have seen how their genuinely held insights and connections lead finally to an objective work situation which seems to be entrapment rather than liberation. But how does this happen? What are the basic determinants of those cultural forms whose tensions, reversals, continuities and final outcomes we have already explored? (p. 119)

Penetration and limitation are concepts which try to come to grips with the paradox not only of why any unfree conditions could be entered freely but also why social change does not ensue from the counter-school's cultural resistance and opposition to the system. Penetration represents those impulses within a specific cultural form that appear to go beyond the confines of the system. I would call penetration the position of the "self as agent." Given that penetration is not simply a characteristic of a person but of a culture, this agency is expressed as persons in relation (i.e., culture). *Limitation* is meant to designate those blocks or impediments that confuse or impede full development and expression. *Partial penetration* is meant to designate the interaction of these two terms in a concrete culture. Partial penetration is what the ethnography reveals, because it describes the field of play in which impulses and limitations combine but cannot be isolated as shown separately (Willis, 1977, p. 119). At the level of pure theory, one could call penetration the "agency of the project" and limitation the "ideology of the habitus." In concrete cultural expressions, agency is always embedded in ideology, as Willis says, *partially penetrated*. Ideally, penetration is the expression of the agency of the personal world in McMurray's terms. Willis puts it similarly:

> I suggest that the smallest discrete unit which acts as the basis of cultural penetration is the informal group. The group is special and more than the sum of its individual parts. It has, in particular, a social dynamic which is relatively independent of issues and locations, preconceptions and prejudice. (p. 123)

The personal world of specific informal cultural forms is the source of *creativity* that holds out the potential for structural change. For example, the counter-school culture succeeds, in part, in demystifying the school system and the working class's lack of options in a class society. This is a cultural insight and not necessarily held by any one individual in the counterculture. The lads' counterculture is potentially creative through their creative resistance to the larger system's work ideology. For example, the lads see through the overemphasis on qualifications, the ludicrous nature of sacrifices to obtain worthless recognition, and the remoteness of real opportunities for upward mobility and intrinsic job satisfaction. The problem with creativity and critical acumen at the level of the informal group is that the system is not exposed at the level on which it operates (i.e., formal–structural). The potential for real creative social transformation and creativity is absent because of the severe limitations operating at the level of the informal group.

Limitation. This word refers to or designates those blocks and divisions that impede the full expression of the impulses directed toward cultural penetration. Willis contends that the two most important divisions that disorganize and divert potential penetration are (1) the divisions of mental and manual labor and (2) gender. To a lesser extent, Willis considers *racism* as another division that disorganizes working-class culture from within. Although the lads stand together as an identifiable subculture, they are divided from other sections of the working class. The lads overvalue manual labor and consider the ear-oles effeminate because of their willingness to be part of a white-collar working force. In this manner, the lads disqualify themselves from the possibilities of being involved with mental labor.

Sexism also disorients cultural penentration. The lads' counterculture promotes its own sexism and celebrates it as part of their overall confidence. As it turns out, their sexism reflects attitudes shared by the wider working-class culture. In calling ear-oles effeminate, one detects implicitly the perjorative attitude toward women. As Willis indicates:

If the currency of femininity were revalued then that of mental work would have to be
too. A member of the counter-school culture can only believe in the effeminancy of
white collar and office work so long as wives, girlfriends and mothers are regarded as
restricted, inferior and incapable of certain things. Despite their greater achievement
and conventional hopes for the future, "ear-oles" and their strategies can be ignored
because the mode of their success can be discredited as passive, mental and lacking
robust masculinity. (pp. 149–150)

Masculinity is seen as both destructive and affirming "labor power." The
masculine disdain for qualification, although housing the insight that mental
work simply helps in the maintenance of class divisions, also allows members of
the middle class to leave the underside of capitalism unquestioned. This lends to
the belief that lower classes are where they are because their mental capacities
keep them where they are. When one considers the dominant ideology and the
value given to different types of labor, it is clear that all classes above the working
class do not confound gender and the manual–mental labor distinction. Further
they celebrate, justify, and see comparatively their superiority in the mental
mode, which is the currency of value in the dominant ideology and culture.

Finally, racism is another division between members of the working class
which is a source of division providing an evident underclass to be looked down
upon by the white working class. It is a source for projecting by the lads of the
degeneracy of others (e.g., West Indian and Pakistani) and the superiority of the
self. Thus, racism divides the working class both materially and ideologically.

What is important in the first part of Willis's analysis is that *penetrations*
and *limitations* are the moments in the analysis which focus on the culture
proper. Penetrations and limitations express the possibilities and limitations for
agency at the cultural level. Willis makes it clear that it is impossible in his study
to delineate the full complexity at the cultural level with respect to outside
deterrents and structure. Nevertheless, he suggests within his own study that it is
possible to suggest one important interface between agency and structure or (as
he put it) between culture and ideology. This brings him to discuss the role of
ideology as part of his analysis, considering the downward impact of structural
class conditions on the counterculture.

Ideology and Culture

The notions of penetration and limitation are concepts which can be considered as the problematic of "personal agency". It is the *agent* side of the
agent–strucure dialectic. In its pure form (i.e., ideal) it is part of the problem of
the projective character of human action. Ideology is the habitus of human
action. Willis acknowledges that his emphasis on cultural penetration and the
internal cultural limitations, if left at that, would underemphasize the impact of
external forces, state institutions, and dominant ideologies acting on the working-class lads. Let it be understood that Willis sees this influence in a dialectical

manner; that is, ideology works on and in the cultural. Nevertheless, one must understand that the agent side of the cultural is formed in the backdrop of dominant ideological systems (i.e., habitus). Exclusive of cultural limitations, ideological systems play an important role in the further limitation of cultural penetrations. Where does the ideological exist? Willis maintains that ideological systems arise partly from ruling or central ideas of the age embodied in concrete institutions (e.g., schools, state institutions) and also the more informal media (i.e., TV, radio, etc.). Willis restricts himself to two core concepts in his discussion of ideology (i.e., structure) and its relation to the cultural (i.e., agent). The addition of notions of agency and structure are my interpretations of the meaning of his concepts. At any rate, Willis pinpoints two salient downward vertical impacts of ideology on countercultural penetrations as those of *confirmation* and *dislocation.*

Confirmation is the solidification of those aspects and resolutions of cultural processes which are congruent with the current social organization and interests of production. Within the school, this ideological confirmation of the present system of production comes through career films and vocational guidance. Willis indicates that the clearest example can be seen in the role differentiation between the sexes in careers films. The films do not offer obvious discrimination and sex-role stereotyping, but the sum total of visual images and the implied assumptions remain sexually divided and are picked up by the sexism of the lads' counterculture:

> In a cake factory we see only girls working and the voice-over tells us as we watch a girl icing a cake, "yes, she really is doing it that quickly, the film isn't speeded up. Women are always doing intricate and fiddly things. Their only relation to the more heroic occupations is one of fear and concern for their men-folk. (Willis, 1977, pp. 161–162)

The role of ideological confirmation allows uncertain cultural resolution (e.g., sexual division of labor) to be considered as part of the nature of things (i.e., naturalism). What it does is inform common sense about the "nature of things."

Dislocation is a process whereby something new is brought into a cultural system and has the ability to mute the partial critical insight of that cultural system. For the lads, solidarity with one another is a very important value and has practical significance. This value of solidarity is dislocated and challenged by the dominant ideology of the school. Solidarity is dislocated by the school's stress on and valuing of individualism. A careers teacher is quoted by Willis as follows:

> I've told you before, it's not often I advocate selfishness but in this case I must. Forget your friends sitting around you now you might be together, now having a laugh, and it doesn't matter. But all the friends in the world are no use to you when it comes to finding a job. . . . So, just this time, be selfish; don't worry what your friends are doing, you get out and look after number one, sort out your own jobs and don't wait for your mates. (p. 164)

The confirmation and dislocation of ideological messages surrounds culture with a sense of pervasive naturalism. When blame is ventured, ideology has a message that maintains the integrity of the social structure: "Human nature, not capitalism, is the trap. Ideology has helped to produce that, though not simply from its own resources—it is believed because it is partly self-made" (p. 165).

The Internal Interlocutor. Willis stresses the main impact of what he is calling "ideology," and specifies that his immediate concern is not its direct intervention on cultural penetrations. The implicit work of ideological dislocations works on the cultural world in the following fashion. The lads celebrate their resistance to the formal as an exception to the rule. There is no cumulative impact in the lads resistance. It proceeds blindly to all other's exceptions which, if summated, could overthrow the rule. This lack of awareness of the real possibility of their resistance is a rule unto itself. The suppressed awareness has vulnerability implicit in its silence. Willis contends that it is in this silence that ideology strides. Willis introduces his own version of the devil in what he calls the "internal interlocutor": "Whether right or wrong, whether penetrated or not, it is *the rule*, it is the *voice*. It becomes the internal interlocutor for the weakness of cultural forms" (p. 166).

The voice (internal interlocutor) is the "them" in "us"; the part of the self that accepts the legitimacy of the system even when resisting this has important political implications, which Willis is at pains to emphasize:

> That the them in us survives in us is usually overlooked. This internal division should not be surprising. In a peaceful social democratic society with real class divisions, the "them" and "us" can never be starkly clear. This basic distinction must be a rehearsal and mediated and echoed around from the largest social units to the individual person. . . . It is this which allows the "us" to properly betray itself. Ideology is the "them" in "us." It has been invited. Informality and the strength of personal validation unconnected with a political practice invite it. . . . Once there it confirms partiality and dislocates penetration. It prevents the "us" from becoming a collective assertive "we." (p. 169)

The absence of a personal world, as we have indicated from the outset, has social and political implications, as stated by Willis above. An authentic personal world is necessary for a political praxis. It is the realm of freedom, because it represents the organization of self-in-relation to the future (Willis, 1977):

> Social agents are not passive hearers of ideology, but active appropriators who reproduce existing structures. Capitalist freedoms are potentially real freedoms and capitalism takes the wager, which is the essence of reproduction, that freedom will be used for self-damnation. (p. 175)

ADEQUACY OF THE ACCOUNTS: A COMPARISON

The critique that I will now make of these studies is, in part, a reflection of my own interpretive horizon which by now is known to the reader. Let us consider the issue of adequacy under the four dimensions that I have previously described at the beginning of this chapter.

1. *An adequate account is negotiated.* The authors of *The Rules of Disorder* have a sensitivity for the boys that they are trying to understand and are thus able to part company with media accounts of these youth; they counter in their own interpretative accounts the media definition of these young boys as simply violent. In their account, they attempt to show order where disorder appears at the surface of their actions. Their discussion of "moral careers" as a way of maintaining a sense of dignity indicates that the authors have a feel for the project character of human actions in this instance. In fact, I would say that the notion of moral careers constitutes a partial project for these boys. Although their effort is not systematic or extensive, the authors do attempt to negotiate their accounts with the participants. There is no extensive coverage in this negotiation, so one must be somewhat critical of them here for not carrying out an extensive negotiation with some detailed summaries. This is important, because the authors venture that their account is valid because it does not fly in the face of the actors' interpretive account of their own actions. In other words, they contend that their account is negotiated with the participants. It is our impression that the negotiation is not well documented or discussed and should not be used, on this basis, as a validating mechanism.

Willis's account exemplifies what I deem to be a critical interpretative account of the personal. It is personal in that it captures in ethnographic form (i.e., first-person accounts) a specific cultural world and its strivings (i.e., projects). It is critical because it systematically links the personal world of culture (ethnography) to the structural dimensions of political economy. It is, to my mind, a systematic attempt that explores the duality of structure in what I have previously called the agent–structure dialectic. It is clear that Willis takes a critical perspective toward his own interpretations (i.e., reflexive). Willis indicates that one of the pitfalls of participant observation is its tendency toward the justification of the status quo (conservatism). There is a tendency with participant observation to move in the direction of naturalism. This was indeed the case with the *Rules of Disorder* account, whose naturalism was couched in the sociobiological concept of aggro. Willis is at pains to avoid this naturalistic drift by stressing the historical constitutive character of the cultural life. Another problem with the participant-observation method is its tendency toward condescension in relation to the participants. Even when it takes a sympathetic attitude toward the participants, it is frequently patronizing (Willis, 1977). He sees this as a real problem for social scientific research and ponders at one point whether it is possible to imagine an ethnographic account that challenges the presupposition of the researchers involved. Willis poses for himself the following concern, which he considers must be the concern for the researcher–interpreter of the cultural world of others:

> Theories must be judged ultimately for the adequacy they display to the understanding
> of the phenomenon they purport to explain—not to themselves. This book has at-
> tempted . . . to take advantages still offered by a qualitative method to respond de-
> scriptively and theoretically to a real and difficult level of social existence whilst
> resisting tendencies toward empiricism, naturalism and objectification of the subject.
> (p. 194)

What follows from the above considerations is an edited transcription of a group discussion by some of the lads who were given the opportunity to read early drafts of his books. What clearly indicates the negotiability of this account with the participants involved is the opportunity during this group discussion to critically evaluate his role as researcher and what the results meant to them as particpants. For the most part, the participants were interested only in the ethnographic section and not the analytic section. As one participant said:

> Well, I started to read it, I started at the very beginning, y'know, I was gonna read as
> much as I could, then I just packed it in, just started readin' the parts about us and
> then little bits in the middle. (p. 195)

The complexity of the analysis section may discourage these boys because the analysis of class structure is couched in language that is foreign to their cultural world. So, negotiation is limited partly by the cognitive world of the participants. Nevertheless, they do speak to the question of veracity of material where the ethnographic accounts are considered. As Chris indicates: "Almost 90 per cent of what I've read in there was, I can actually remember" (p. 196).

In negotiating the account with the participants, one also can sense the importance of establishing the *credibility* of the researcher. With these lads, it is clear that it is not established by professional credentials. As Joey is quick to point out:

> When you first started asking questions, something illegal must have come out and
> we'd told you things we'd done wrong and we never got any backlash off other
> members of staff which obviously meant you hadn't told anybody. (p. 197)

What is clearly evident from this quote is the trust that must be established if the participants are to reveal themselves. What is clear from this is that there are no "social facts" just there to be collected. They are revealed for interpretation when the researcher is assumed to be credible.

2. *An adequate account is an argument.* By argument, we mean that a particular interpretative account is seen as an advocacy for a particular interpretative horizon with more or less plausibility. In *The Rules of Disorder* the authors understand their account as one that is backed by argument. In fact, their argument attempts to counter media interpretations of youth violence as arbitrary and anarchic. They venture their own argument at two levels. The first is that of a structural articulation of rules interpreting violence at the school and football games of working-class youth. Here the concept of moral careers was introduced

and is ventured as a reason for action. They talk about moral careers as conventional accounts. They then shift to what they say is a causal account embedded in naturalistic explanation. This is their conception of aggro as ritualized aggression. This sociobiological concept is presented as a causal mechanism for violence in these settings. The argument for structural and functional explanation appears to me, however, to be gratuitous. They do not argue in any convincing manner why both must be present as an adequate account. The aggro concept appears as a rabbit out of the hat. Its pretense in the account of human action shows the interpretation in such a way as to prevent this account from being grounds for an emancipatory praxis. The authors do not clearly indicate here their reasons for venturing into sociobiology. Given that they use the concept as a causal mechanism, their thin line of argument for an organic metaphor leaves serious holes in their argument, which I will deal with shortly.

Willis's work, as an account, is seen by the author and revealed to the reader as a particular interpretation of social facts. Recall that the major setting of this study is the institution of the school. Within this work, Willis argues against reproduction theories that ignore the active role of culture in both the maintenance and transformation of a particular society. Specifically, they are in conversation with two major theoretical formulations. The first is the work of Bowles and Gintis, *Schooling in Capitalist America* (1976), and the second is Bourdieu and Passeron's book, *Reproduction in Education, Society and Culture* (1977). It is from this latter work that I have taken the term *habitus*. What is glaringly absent from this work is a notion of *cultural project* which sets up the dialectic of *habitus–project* that I developed in Chapters 3 and 4. With only a notion of habitus, Bourdieu and Passeron (1977) end up by placing sole emphasis on structural factors in the reproduction of cultural learning; the human agent appears hapless before reproductive structural totalities. The power of institutions is conceptualized as "symbolic violence." For them, the school is, par excellence, the structural institution for the reproduction of culture in the young. The educational system inculcates the dominant class position through different cultural constraints. There are different power relations in school according to whether the school caters to elites or groups lower on the status hierarchy. Their argument, as well as that of Bowles and Gintis (1976), accents the pervasive power of economic structures and their linkages to the structure of the school. One could call them structuralists in this light. What is missing is the real or potential role of the agent within culture to reproduce or possibly transform structures that influence their lives. Their structuralism leads a total pessimism toward the resistance or potential projects of the human agent. Their emphasis is on habitus to the exclusion of agency. The school inculcates different classes and the process of inculcation is internalized. Habitus is capable of perpetuating itself after pedagogic action has ceased.

Willis enters here with a concept of culture analyzed through ethnographic first-person accounts. Within this account, the concepts of penetration and limitation are set up to enter into a dialectical relationship with what Bourdieu and Passeron (1977) call "the symbolic violence of structures." The subcultural world of the lads, then, is the focal point not only for the reproduction and continuity of structures but also for their potential transformation. Introducing the agency of subculture in his emphasis on resistance and penetration, Willis opens the door to a realm of freedom which is absent in the other theoretical formulations. He does this without underestimating the powerful determining forces of economic structures which are represented symbolically in the school.

The argument of Willis opens the door to a notion of project and agency where it was heretofore absent. His emphasis on culture and its dialectical relationship to the class structure of modern society under monopoly capitalism would classify him as a cultural Marxist within a Marxist formulation. The argument that he ventures also implicitly critiques formulations that exclude the notions of personal agency. When he argues against naturalism, I take that to include *The Rules of Disorder*. Here, personal agency collapses into a sociobiological formulation called aggro which leaves the economic structures operating in society and selectivity on agents of different social classes unquestioned and uncriticized. Aggro takes what Willis calls "penetration" and casts it in a biological organismic light which I would label "psychologism." Psychologism is where the agency–structure dialectic is collapsed into the psyche. To be sure, it must be understood that it is not a foregone conclusion that one of these interpretations is transparently more adequate than others. The adequacy must be argued because they are conflicts of interpretations in what are apparently similar phenomena. This is why I deem that one criterion of adequacy is that the theory must present itself as an argument to be evaluated rather than a truth that appears to be unassailable.

3. *An adequate account expresses an emancipatory praxis.* In the third criterion of adequacy, I have maintained that an account must express an emancipatory praxis. It makes a case for the freedom of agency. *The Rules of Disorder* contradicts the possibility of freedom in the aggro concept. In interpreting order in this manner, the authors limit the scope of the project exemplified in the development of the moral careers. One must ask here why dignity must fall to a ritual that locks people into institutions that prevent them from transforming their lives. Why does this ritualized aggression seem to exercise itself in a political social situation? The question of the class manifestation of a ritual is never asked. Aggro and its biological gloss explains why things are the way they are. It has no openness to a transformed future where myth has utopian rather than simply ideological elements. If Job were to be confronted in the 20th century, possibly these authors would be one group of his comforters.

It must be understood that the authors of *The Rules of Disorder* would not judge themselves under this norm. In fact, the *value* of emancipation is one that most social psychologists tend to ignore. One chooses it as a norm for adequacy if, in one's theory, there is an interest in human freedom and emancipation. Specifically, I have outlined two minimum conditions: (1) The account must show how the immediate social and material conditions that determine the social world are mediated through the agency of actors who have some control over the constitution of their world. In Willis's study, this is precisely the importance of the ethnographic data and the corresponding concepts of penetration and limitation. The reader senses in the accounts the personal responsibility of actors in the partial creation of their world. (2) The account will show how taken-for-granted self-understandings are socially mediated and could be otherwise (i.e., indicates alternatives). Willis resists a naturalism (e.g., aggro) in his accounts, and his formulation of penetration indicates the possibility for structural change. His critical analysis of limitations in the lads culture (i.e., sexism and racism) indicates that he does not see these inevitable outcomes of class struggle. In other words, he sees the lads as having the capability of intentional intervention into these structures, that is, the capacity of power to change them (i.e., transform). In other words, lower-class boys do not have to be sexist. In introducing the possibility for intentional intervention with his notion of subcultural penetration and resistance, he nevertheless is soberly aware of the limitations that hamper the full impact of personal agency. He outlines and discusses these structurally limiting factors under the terms *racism, sexism,* and *individualism.* All these limiting factors detract from a solidarity potent enough to challenge class structural totalities. Because Willis's interpretative account has a concept of intentional intervention in the duality of penetration and limitation, his account expresses, to my mind, an emancipatory praxis. It should also be noted that he gives some specific suggestions for a different type of vocational guidance and a different type of education for disaffected working-class youth. Part of a long chapter deals with issues he thinks are important for teachers and vocational guidance personnel who have occasion to work with cultures like that of the lads. In recognizing the limitations of the lads' culture, he offsets this with an appreciation of the partial penetration of these youths who criticize the larger culture. He is thus prevented from the all too common drift toward blaming the victim.

    4. An adequate account is critical. Finally, an adequate account must be a critical interpretation of the world of personal agency. As was said in the previous chapter, critical interpretation as a way of resymbolizing the world of others requires a sensitivity in reflecting back to the participants and other interested parties (e.g., social scientists, social policymakers, etc.), their capacities for *intentional intervention* (i.e., free acts of agents in creating a world). *The Rules of Disorder* clearly lacks this critical moment as an account. As we said earlier, a critical account must resymbolize the world of others and should sensitively reflect back to them their capacities for intentional intervention. We see a class bias in this interpretation, since the aggro ritual simply leaves the actors in their state with a myth to live by. The account seems to ignore the agent quality of aggression and the possible truth that rebellion is more than a ritual liability. The authors interpret resistance as a biological process rather than one form of intentional intervention, however unsuccessful to arbitrary indignities produced by the class structure in modern societies.

By contrast, *Learning to Labour* presents the idea of intentional intervention by using the concept of partial penetration. The whole idea of partial penetration, as seen in the lads countercultural resistance, is indicative of the possibilities for intentional intervention in the world of these lads. What is different about Willis's account of lower-class rebellion and resistance is its positive value within his interpretive system. In contrast to much social science interpretation of lower-class strata, he does not move to interpret their rebellion as simple deviance. At the same time, he is able to state some real reservations about "resistance" within the context of lower-class penetration. Here a critical interpretation of personal agency also considers the liabilities of a particular cultural form. This is clearly seen in Willis's extensive discussion of what he calls "limitations" and what I am calling "liabilities." What is unique about Willis's account is that he treats liabilities dialectically in relation to agency (i.e., penetration) and embeds the agency–liability dialectic into the agency–structure dialectic. In presenting liabilities within the agency–structure dialectic, he does not fall into blaming the victim for all the liabilities to personal agency. At the same time, this agency–structure dialectic allows him to attribute some personal responsibility to the agents involved, with full knowledge of the powerful effects of social structure on the possibilities for personal agency.

To this point, I have presented the most positive features of the study as I understand it. This I have done for two reasons. First, I think it is an excellent study, barring some reservations that I am about to discuss. Second, the study exemplifies all the significant features and concerns that I have developed throughout this work as a critical interpretation of the personal world. It should be pointed out that Willis does not present himself as a psychologist. So why call this study a psychology that exemplifies a psychology as critical interpretation of the personal world? I do this because psychology has traditionally laid claim to the world of the microcosm, be it individual or person. In that sense, Willis is incorporating—by use of ethnography—the interpersonal world of persons in relation (i.e., culture) within the context of a social structural class analysis of a Marxist variety. One, therefore, sees this study as a nascent form of inquiry which is interdisciplinary in nature. This is exactly what psychology as a critical interpretation is all about. Therefore, psychologists who attempt a critical interpretation of the personal world will need a greater knowledge of cognate disciplines such as anthropology, sociology, political science, and economics. In the past, the socialization of a psychologist into other disciplines was reductionistic in orientation, knowledge of biology and chemistry and physics being the main sister disciplines. This is not to say that psychologists must abandon these latter goals. What is being said is that it must include the former disciplines if the psychology being propagated is to have a critical intent.

Let me now conclude with some caveats on Willis's study, so the reader will understand that no study, critical or uncritical, is above criticism. Some caveats are as follows. Ethnography is labor-intensive work, so it is understandable that samples exemplifying a culture will be small in number. Nevertheless, this study never addresses the problem of its small sample size or the fact that there are a great deal of narrative data on only a few of the lads. Joey seems to be extensively quoted, whereas others are not mentioned. I think Willis has not adequately addressed this problem and the corresponding issue of universality and generality that is drawn from a small sample of a subculture. To what extent does a small sample of this nature represent a subculture? This question is not adequately addressed in this work. One can also say that he makes too much of lower-class resistance and rebellion. Given the apolitical nature of their rebellion, which the author freely acknowledges, one could ponder their rebellion as "alienated" rebellion and opposed to "integrated" rebellion. Here I am continuing a distinction that I made in Chapter 4, when I discussed the Watts uprising. Resistance is easily classified as deviance when rebellion is not centered on a project with *utopian* symbols. The lads have no utopian symbols that project beyond their present state in the class structure. This is evident in the quotes supplied by Willis. Recall that when Watts reached the critical juncture of an integrated black cultural project, rebellion and resistance were coupled with utopian goals which led to cultural projects in the ghetto that even detractors had to acclaim. The lads macho, individualism, sexism, and racism are hardly the stuff of social transformation and utopian striving. All movements of social transformation (violent and nonviolent) have a moral persuasion that not only challenges the enclaves of the status quo but also their moral integrity. This is demonstrated in the utopian and charismatic leaders of all modern movements of social transformation (e.g., Gandhi, Cesar Chavez, Malcolm X, Martin Luther King). I am not convinced by Willis that "cultural resistance," which the lads demonstrate, can effectively transform the class structure by simply being politicized. The power of their project is not only weakened by their own divisiveness but also the total absence of real utopian symbols. Without engaging in victim blaming, one must nevertheless see their rebellion in the present state as alienated in nature (see Table 4). This alienation must not be reduced to a sociobiological concept like aggro. It must be seen as *historical* in nature and capable of being transformed under the right social condition by utopian projects.

METHOD REVISITED

The two stuides just described in their particularity can now be seen in a more generic and historical context. Both studies strike a very different balance on the agency–structure dialectic. Their differences, however, have a long social and political history to explain why they essentially come down on opposite poles of an agency–structure dialectic. Dawe (1978) traces this to a fundamental dualism which has emerged historically from the Enlightenment formulation. He elucidates the dualism as follows:

> There has been a manifest conflict between two types of social analysis, variously labelled as being between the organismic and mechanistic approaches, methodological collectivism and individualism, holism and atomism, the conservative and emancipatory perspectives, and so on. The debate about these issues . . . are all different versions of the fundamental debate about the abiding conflict between the domination of the system and the exertion of human agency. (p. 366)

Dawe contends that the basic contemporary conflicts in the human sciences are around these two opposing views of human potential. Theories that accent the importance of the social system are willy-nilly pessimistic about human actors, envisioning them as manipulable but as also tending on their own initiatives toward destructive actions. The bottom line of the above perspective sees the survival of society in the containment of human agency. The essential emphasis is on social order (i.e., habitus).

The other side of this dualism emphasizes *freedom* over *order*. Here, the recognition of constraints on human actors is seen as coming from the actions of others. I have referred to this as "domination." Within this perspective, society is the outcome of autonomous, creative human actions. The moral focus of this perspective is the idea of human control or direction over the system (Dawe, 1978).

As a dualism, both systems are inadequate and have severe limitations at their extremes. One system sacrifices *freedom* for the sake of social order, the other sacrifices social coherence for the sake of freedom (i.e., anarchy). Most of what constitutes conventional psychology errs on the side of order. Thus, my contention earlier that psychology serves the cause of the social order over the individual. Even where there is a focus on the individual, conventional psychology implicitly brings the individual in line with the social order. At the same time, most reactions to social order maintenance have been libertarian in nature. I count most of the human potential movement of the 1960s under this aegis.

What is needed now in social theory and practice are perspectives which pose the agency–structure issue in dialectical terms. To my mind, the Willis study captures this dialectical spirit, whereas *The Rules of Disorder* collapses agency into a functional sociobiology which is "social order"-maintaining. But it must be understood that the notion of agent freedom located in our agency-structure dialectic must encompass a freedom that goes beyond libertarianism. Berlin (1969), in a discussion of two concepts of liberty, calls the concern for freedom that I am involved with status *freedom. It is not individual liberty* in the traditional liberal sense. Rather:

> It is something no less profoundly needed and passionately fought for by human beings—it is something akin to, but not itself freedom: although it entails negative freedom for the entire group, it is more closely related to solidarity, fraternity, mutual understanding, need for association on equal terms. . . . The desire for recognition is a desire for something different: for union, closer understanding, integration of interests, a life of common dependence and common sacrifice. (Berlin, 1969, p. 158)

I differ with Berlin (1969) that one can associate liberty from solidarity. This disassociation has been characteristic of the liberal concept of freedom. But freedom as allied with *solidarity* is achieved through *totalitarianism*. It must be clearly understood that the sense of freedom achieved under these conditions is a caricature of the human efforts toward creative agency.

To conclude, the belief in personal agency, project, freedom, and emancipation, which I have engaged the reader in through this book, is a *value* rather than a *fact*. As a *value*, these interests must come under *factual* scrutiny. In other words, one can ask: Are these values real? The answer to this question for a professional psychologist such as myself will not come from within my discipline. These concerns, which are moral in nature, one brings to the discipline. I fully understand that to say that psychology is the study of the personal world of human freedom flies in the face of much of what constitutes the discipline of psychology. I can only hope that my ideas have been developed in this book to make the idea of freedom a tolerable idea for some. I am, however, more ambitious than that. I would hope that an interest in *human emancipation* would strike a chord which is cordial rather than barely tolerable.

It seems to me that an interest in human freedom is more vital today than at any other period in history. If one looks at Skinner, one gets the impression that "freedom" has failed: "Now let's try some systematic control." It is not that freedom has failed. In relation to freedom, one can repeat Chesterton's words on Christianity: "It has not been tried and found wanting, it has been found wanting but hasn't been tried." This certainly can be said of psychology in its short history.

Appendix

I have extensively discussed two studies in Chapter 7 to illustrate, in detail, what a critical psychology might look like and what it is not. It is important for the reader to understand that the Willis study is one minute example of what a critical psychology might look like. Other studies may be, and are, radically different from this one yet can be considered critical in intent. What is important is that the study incorporate the features of a critical study that I have developed in this book.

My major critical concerns in this book were directed toward structures of domination based on class differences in economic opportunity, sexism, racism, and adultocentrism. Other critical theories may not have the same focal concerns. In fact, one can analyze some studies that do not issue from a critical theoretical orientation and consider some of the critical features that these studies exemplify. This is important at first, since there are very few psychological studies having an explicit critical orientation. I would like to conclude by giving several examples to round out my treatment.

1. *Let Us Now Praise Famous Men.* This work by James Agee with photographs by Walker Evans focuses on a poor white family living in the American South during the 1930s (Agee & Evans, 1974). It is an intensive study of their personal world and also depicts how the researcher–writer (Agee) was affected by his stay with them. Instead of blaming the victims, Agee attempts to show by detailed description how people work and live under oppressive social conditions. While not sidestepping the oppression these people suffer, he nevertheless establishes their tremendous power in working together. This is a very moving ethnography of the personal world of the participants and the researchers. It brings to the fore the best of liberal sympathies. What is absent in this ethnography is the relation of the personal cultural world to the larger economic conditions which enforce poverty as a way of life among these people. It is, thus, lacking the agency–structure dialectic.

2. *Darwin on Man.* This exploration of scientific creativity by Howard Gruber (1974) is an intensive psychological study of Darwin's formative development that views his discovery as a communal enterprise. Particularly, this study brings out the notion of project or enterprise discussed in an earlier chapter. What is clear from this psychohistory is the relation of Darwin's lifelong project to the people around him. Darwin's personal agency is agency in relation. The notebooks indicate the tremendous personal support that Darwin received from the women who surrounded him. One never gets the impression from biology textbooks that the theory of evolution also owes something to women,

177

who gave up their project for its author. Gruber's study is important because it dramatizes the notion a project within the context of creative scientific theorizing.

3. *The Privileged Ones.* This work by Robert Coles (1977) differs from most of the studies discussed in this book, which have focused on the working class or racial underclasses. This approach, however, represents only a partial understanding of the ethnography of class. If Willis tries to comprehend why lower-class children come to accept working-class jobs, Coles takes the middle-class child and tries to understand why "children cultured in affluence" come to see themselves as deserving wealth and luxury. This extensive ethnography of affluent children attempts to understand the development of the "project of affluence" (i.e., I expect to be as rich as or richer than my parents) and outlines the concept of "entitlement," which is a pervasive psychological expectation that things must go your way (i.e., no resistance is expected). Coles's study, therefore, gives one an insight into how upper-class children are socialized into power roles in our society. To use a Hegelian phrase, the study helps us understand the master in the master-slave relation. What is totally absent is the agency–structure dialectic, which would make it a critical account.

4. *The Hidden Injuries of Class.* This study by Richard Senett and Jonathon Cobb (1973) seeks to understand how the class structure and nature of work affect the lives of adult working-class males. It shows how class structure is reproduced in the personal lives of these men. Gone is the rebellion and resistance that seem to characterize their youthful counterparts in the Willis study. What seems to be indicated is that the lads will eventually be subdued by the class system by disintegrating working conditions. What is dramatized here is not the presence of personal agency and cultural solidarity but the full working out of "alienated labor" and the presence of psychological liabilities that make these men accept the system and their lives within it as inevitable. The victims blame themselves. One senses here how the structural conditions of capitalism, and the alienated labor that Marx claims ensues, affect and subdue the lives of working-class men. They "buy the system" without benefiting from it. I am suspicious of the pervasive pessimism of this study. It seems that there is no resistance in these males, and yet Stanley Aronowitz, in his *False Promises* (1973), makes it very clear that there is a considerable amount of resistance in the work force that shows up in subtle ways. In this, Aronowitz sides with Willis, risking to err on the side of optimism.

5. *World of Pain: Life in the Working-Class Family.* This work by Lillian B. Rubin (1976) is an intensive study of a group of white lower-middle-class families. Based on interviews with men and women who make up these working-class families, the study aims to link the interiority of family life (i.e., microcosm of a personal world) with the work that these people do. In other words, it assesses their place in the class structure and the effect it has on their personal lives within the family. As the author states herself, "Indeed, while there are hundreds of studies of marriage and the family each year, few even mention the word "class"—an indication of how invisible the class structure has been to those who conceptualize and carry out those projects" (Rubin, 1976, p. 5).

What is dramatized in this study is how the family of working-class individuals are subdued by the cultural and economic system. This study, like the previous one, captures the limitations and liabilities of personal lives under capitalism. What seems totally absent is the possibility of "personal project," which holds out the possibility of social transformation.

Like the Senett and Cobb study, it appears to characterize the total victory of ideology (habitus) over utopia (project). In spite of its obvious merits and sensitivities, I hold it suspect for its total eclipse of projective quality of human action.

6. *The Life of the Self.* Robert Jay Lifton's study (1976) concerns American soldiers who turned against the war they were fighting (Vietnam). There are many interesting things about this study, but most important for our purpose is how the researcher assumes an advocate role for the people he is studying. By *advocate*, I mean that this psychiatrist shares the moral rejection of the war that his participants express. He purposefully resists calling his sessions with these veterans "psychotherapy" because he considers their war guilt a moral sensitivity rather than the focus for psychiatric diagnosis. He calls his group a "rap group" and he intentionally attempts to enter into a reflexive–dialogical relation with the participants. He assumes a critical stance toward his own profession of psychiatry, which is aiding the cause of an immoral war by treating resistance to this war by some veterans as "deviance."

7. *Sane Asylum.* In this work, Charles Hampden-Turner (1976) studies a community of reformed addicts and alcoholics. The study is an *ethnography* with no discussion of the effects of social structure outside of the small groups considered. Nevertheless, it focuses on a microcosm of seemingly hopeless and wasted people and shows how, through a certain type of communal living, their lives are transformed from almost total "liability" to "project". In this study, one gets a feel for the notion of interdependence, which is what I mean by the enhancing power of a personal world. If the study errs, it will do so on the side of optimism. It captures the "utopian symbolism" released by human solidarity and shows how it can transform seemingly hopeless lives. One really gets the impression of a movement from oppression to liberty in the lives of a specific group of people. For all of its shortcomings, which I have failed to discuss, its interest and ability in capturing human emancipation is laudatory.

I have just given a small sampling of studies as examples to help orient the reader to a world of critical interpretation. The reader must now take the initiative to provide further examples. My selection reflects some of my own idiosyncrasies.

BIBLIOGRAPHY

Abel, T. The operation called Verstehen. *American Journal of Sociology*, 1948, 54, 211–218.

Agee, J., & Evans, W. *Let us now praise famous men*. New York: Ballantine Books, 1974.

Althusser, L. Ideology and the state. In *Lenin and philosophy and other essays*. London: New Left Books, 1971.

Aries, P. *Centuries of childhood*. New York: Vintage, 1962.

Aronowitz, S. *False promises*. New York: McGraw-Hill, 1973.

Ausubel, D. P., & Sullivan, E. V. *Theory and problems of child development* (2nd ed.). New York: Grune & Stratton, 1970.

Ausubel, D. P., Sullivan, E. V., & Ives, W. S. *Theory and problems of child development*. (3rd ed.). New York: Grune & Stratton, 1980.

Bakan, D. *The duality of human existence*. Chicago: Rand McNally, 1966.

Baum, G. *Religion and alienation*. Toronto: Paulist Press, 1975.

Bauman, Z. *Hermeneutics and social science*. New York: Columbia University Press, 1978.

Becker, E. *Escape from evil*. New York: Free Press–Macmillan, 1975.

Belaval, Y. Vico and anti-Cartesianism. In G. Tagliacozzo (Ed.), *Giambattista Vico: An international symposium*. Baltimore: Johns Hopkins University Press, 1969.

Berlin, I. *Four essays on liberty*. New York: Oxford University Press, 1969.

Berlin, I. *Vico and Herder: Two studies in the history of ideas*. New York: Viking, 1976.

Berlyne, D. E. *Structure and direction in thinking*. New York: Wiley, 1966.

Bernstein, D. *Class, codes and control*. London: Routledge, 1975.

Bernstein, R. *Praxis and action*. Philadelphia: University of Pennsylvania Press, 1971.

Bernstein, R. J. *The restructuring of social and political theory*. Philadelphia: University of Pennsylvania Press, 1978.

Black, M. *Models and metaphors: Studies in language and philosophy*. Ithaca, N.Y.: Cornell University Press, 1962.

Bleich, D. *Subjective criticism*. Baltimore: John Hopkins University Press, 1978.

Bourdieu, T. Outline of a theory of practice. In *Cambridge studies in social anthropology*. London: Cambridge University Press, 1977.

Bourdieu, T., & Passeron, J. C. *Reproduction in education, society and culture*. Beverly Hills, Calif.: Sage, 1977.

Bowles, S., & Gintis, H. *Schooling in capitalist America: Education reform and the contradictions of economic life.* New York: Basic Books, 1976.

Braverman, H. *Labor and monopoly capital.* New York: Monthly Review Press, 1974.

Brenner, H. *Mental illness and the economy.* Cambridge, Mass.: Harvard University Press, 1973.

Brenner, H. *Estimating the social costs of national economic policy: Implications for mental and physical health and criminal aggression.* Washington, D.C.: U.S. Government Printing Office, 1976.

Bronfenbrenner, U. *The ecology of human development.* Cambridge, Mass.: Harvard University Press, 1979.

Broughton, J. Developmental structuralism: without self, without history. In H. K. Betz (Ed.), *Recent approaches to the social sciences.* Calgary, Alberta, Canada: University of Calgary Press, 1979.

Brown, A. L. Knowing when, where and how to remember: A problem of meta cognition. In R. Glaser (Ed.), *Advances in instructional psychology* (Vol. 1). Hillsdale, N.J.: Erlbaum, 1978.

Brown, B. *Marx, Freud, and the critique of everyday life: Toward a permanent cultural revolution.* New York: Monthly Review Press, 1973.

Buber, M. *I and thou.* New York: Scribner, 1970.

Buck-Morse, S. Socioeconomic bias in Piaget's theory and its implications for cross-cultural studies. *Human Development,* 1975, *18,* 35–49.

Buss, A. R. The emerging field of the sociology of psychological knowledge. *American Psychologist,* 1975, *30,* 988–1002.

Cagan, E. Individualism, collectivism, and radical educational reform. *Harvard Educational Review,* 1978, *48,* 227–266.

Campbell, D. On the conflicts between biological and social evolution and between psychology and moral tradition. *American Psychologist,* 1975, *30,* 1103–1126.

Caplan, A. L. *The sociobiology debate.* New York: Harper & Row, 1978.

Chein, I. *The science of behaviour and the image of man.* New York: Basic Books, 1972.

Chodorow, M. *The reproduction of mothering: Psychoanalysis and the sociology of gender.* Berkeley: University of California Press, 1978.

Chomsky, N. *Language and mind.* New York: Harcourt, Brace, & World, 1968.

Chomsky, N. *Problems of knowledge and freedom.* New York: Random House, 1971.

Coles, R. *Privileged ones.* Boston: Little, Brown, 1977.

Connell, R. W. *Ruling class, ruling culture.* London: Cambridge University Press, 1977.

Cornforth, M. *Materialism and the dialectical method.* New York: International Publishers, 1971.

Dagenais, J. *Models of man: A phenomenological critique of some paradigms in the human sciences.* The Hague: Nijhoff, 1972.

Danziger, K. The social origins of modern psychology. In A. Buss (Ed.), *Positivist sociology in the context of psychological theory.* New York: Irvington, 1978.

Davis, G. *Childhood and history in America.* New York: Psychohistory Press, 1976.

Dawe, A. Theories of social action. In T. Bottomore & R. Nisbet (Eds.), *A history of sociological analysis.* New York: Basic Books, 1978.

De Charms, R. *Enhancing motivation: Change in the classroom.* New York: Wiley, 1976.

De Lone, R. *Small futures.* New York: Harcourt Brace Jovanovich, 1979.

De Mause, L. The evolution of childhood. In L. De Mause (Ed.), *The History of childhood.* New York: Psychohistory Press, 1974.

Descartes, R. *Discourse on method.* Baltimore: Penguin Books, 1960. (Originally published, 1637.)

Dickason, A. Anatomy and destiny: The role of biology in Plato, as Plato's views on women. In C. C. Gould & M. W. Wartofsky (Eds.), *Women in philosophy: Toward a theory of liberation.* New York: Putnam, 1976.

Dyer, K. The dialogic of ethnology. *Dialectical Anthropology,* 1979, *14,* 205–224.

Edwards, R. *Contested terrain.* New York: Basic Books, 1979.

Eichler, M. The origins of sex inequality. A comparison and critique of different theories and their implications for social policy. *Women's Studies International Quarterly,* 1979, *2,* 329–346.

Eichler, M. *The double standard: A feminist critique of feminist social science.* London: Croom Helm, 1980.

Erikson, E. Identity in the life cycle. *Psychological Issues,* 1959, *1*(1).

Ewen, S. *Captains of consciousness.* New York: McGraw-Hill, 1976.

Favreau, O. E. Sex bias in psychological research. *Canadian Psychological Review,* 1977, *18,* 56–65.

Fay, B. *Social theory and political practice.* London: G. Allen, 1975.

Fowler, J. W. Faith, liberation and human development. *The Foundation.* Atlanta: Gammon Theological Seminary, Spring 1974.

Freire, P. *Cultural action for freedom.* Baltimore: Penguin Books, 1974.

Gadamer, H. G. Hermeneutics and social science. *Cultural Hermeneutics.* 1975, *2*(4), 307–316. (a)

Gadamer, H. G. *Truth and method.* New York: Seabury, 1975. (b)

Gadlin, H., & Ingle, G. Through the one-way mirror: The limits of experimental self-reflection. *American Psychologist,* 1975, *30,* 1003–1009.

Gagné, R. M. Contributions of learning to human development. *Psychological Review,* 1968, *75,* 177–191.

Gardner, H. *The quest for mind.* New York: Random House, 1972.

Gauld, A., & Shotter, J. *Human action and its psychological investigation.* London: Routledge, 1977.

Geertz, C. *The interpretation of culture.* New York: Basic Books, 1973.

Gerth, H., & Mills, C. W. *Character and social structure: The psychology of social institutions.* New York: Harcourt Brace Jovanovich, 1964.

Giddens, A. *New rules of sociological method.* London: Hutchinson University Library, 1977.

Giddens, A. *Central problems in social theory: Action, structure and contradiction in social analysis.* London: Macmillan, 1979.

Giorgi, A. *Psychology as a human science.* New York: Harper & Row, 1970.

Giorgi, A., Fischer, C., & Murray, E. *Phenomenological psychology* (Vol. 2). Pittsburgh: Duquesne University Press, 1975.

Glasgow, D. G. *The black underclass.* San Francisco: Jossey-Bass, 1980.

Gould, C. C. Philosophy of liberation and the liberation of philosophy. In C. C. Gould

& M. W. Wartofsky (Eds.), *Women and philosophy: Toward a theory of liberation.* New York: Putnam, 1976.

Gramsci, A. In P. Hoare & G. Smith (Eds. and trans.) *Selections from prison notebooks.* New York: International Publishers, 1971.

Gruber, H. E. *Darwin on man: A psychological study of scientific creativity.* New York: Dutton, 1974.

Gutierrez, G. *A theology of liberation.* New York: Orbis, 1973.

Habermas, J. *Toward a rational society.* Boston: Beacon, 1970.

Habermas, J. *Knowledge and human interests.* Boston: Beacon, 1972.

Habermas, J. *Theory and practice.* Boston: Beacon, 1974.

Habermas, J. *Communications and the evolution of society.* Boston: Beacon, 1979.

Hall, D. L. Biology, sex, hormones and sexism in the 1920s. In C. C. Gould & M. W. Wartofsky (Eds.), *Women in philosophy: Toward a theory of liberation.* New York: Putnam, 1976.

Hampden-Turner, C. *Radical man.* New York: Doubleday, 1971.

Hampden-Turner, C. *Sane asylum.* San Francisco: San Francisco Book Co., 1976.

Harré, R., & Secord, P. E. *The explanation of social behaviour.* Oxford, England: Blackwell, 1972.

Hart, H. L. A. Description of responsibility and rights. In A. C. N. Flew (Ed.), *Logic and language,* first series. Oxford, England: Blackwell, 1951.

Hilgard, E. R., & Bower, G. H. *Theories of learning* (3rd ed.). New York: Appleton Century Crofts, 1966.

Hirsch, E. D. *The aims of interpretation.* Chicago: University of Chicago Press, 1976. (a)

Hirsch, E. D. *Validity in interpretation.* New Haven, Conn.: Yale University Press, 1976. (b)

Hodges, H. W. Vico and Dilthey. In G. Tagliacozzo (Ed.), *Gaimbattista Vico: An international symposium.* Baltimore: Johns Hopkins University Press, 1969.

Horner, M. The motive to avoid success and changing aspirations of college women. In J. Bardwick (Ed.), *Readings on the psychology of women.* New York: Harper & Row, 1970.

Hudson, L. *The cult of the fact.* London: Jonathan Cape, 1972.

Hull, C. *Principles of behavior.* New York: Appleton Century Crofts, 1943.

Hunt, D., & Sullivan, E. V. *Between psychology and education.* Hinsdale, Ill.: Dryden, 1974.

Hunt, J. Mc. V. *Intelligence and experience.* New York: Ronald, 1961.

Ingleby, D. Psychology of child psychology. In M. P. M. Richards (Ed.), *The integration of a child into a social world.* London: Cambridge University Press, 1974.

Jacoby, R. *Social amnesia.* Boston: Beacon, 1975.

James, W. *Varieties of religious experience.* New York: Modern Library, 1936. (Originally published, 1902.)

Jameson, F. *The prison house of language.* Princeton, N.J.: Princeton University Press, 1972.

Jensen, R. How much can we boost IQ scholastic achievements? *Harvard Educational Review,* 1969, 39, 1–123.

Kamerman, S. B. Work in family in industrialized societies. *Science—Journal of Women in Culture and Society,* 1979, 4, 632–650.

Kamin, L. *The science and politics of I.Q.* Hillsdale, N.J.: Erlbaum, 1974.

Kellner, D. Ideology, marxism and advanced capitalism. *Socialist Review,* 1978, 8(6), 57–58.

Keniston, A. *Young radicals.* New York: Harcourt, Brace, & World, 1968.

Kessen, W. J. Commentary in insight of the child development movement in the United States. *Monographs of the Society for Research in Child Development,* 1975, *40,* Serial #161.

Kuhn, T. *The structure of scientific revolutions.* Chicago: University of Chicago Press, 1962.

Laing, R. D., & Cooper, D. C. *Reason and violence.* London: Tavistock Publications, 1964.

Lifton, R. J. *The life of the self: Toward a new psychology.* New York: Simon & Schuster, 1976.

Lock, A. *Action, gesture and symbol.* New York: Academic Press, 1978.

Lonergan, B. *Insight.* London: Longmans, 1957.

Lonergan, B. *Method of theology.* New York: Herder & Herder, 1972.

Maccoby, M. *The gamesman.* New York: Simon & Schuster, 1976.

MacMurray, J. *The self as agent.* London: Faber & Faber, 1957.

MacMurray, J. *Persons in relation.* London: Faber & Faber, 1961.

McCarthy, T. *The critical theory of Jurgen Habermas.* Cambridge, Mass.: M.I.T. Press, 1978.

Mannheim, K. Conservative thought. In P. Kecskemeti (Ed.), *Essays on sociology and social psychology.* London: Routledge & Kegan Paul, 1953. (a)

Mannheim, K. *Ideology and utopia.* New York: Harcourt, Brace and World, 1953. (b)

Markovic, M. Woman's liberation and human emancipation. In C. Gould & M. Wartofsky (Eds.), *Women and philosophy: Toward a theory of liberation.* New York: Putnam, 1976.

Marsh, P., Rosser, E., & Harré, R. *The rules of disorder.* London: Routledge, 1978.

Marx, K. *Early writings* (T. Baltimore, Ed.). New York: McGraw-Hill, 1963. (Originally published, 1844.)

Marx, K., & Engels, F. *The German ideology.* New York: International Publishers, 1960.

McNeill, D. Developmental psycholinguistics. In F. Smith & G. A. Miller (Eds.), *The genesis of language.* Cambridge, Mass.: M.I.T. Press, 1966.

Mead, G. H. *A social psychology: Selected papers.* Chicago: University of Chicago Press, 1914.

Merleau-Ponty, M. *The structure of behavior.* Boston: Beacon Press, 1963.

Mills, C. W. *The sociological imagination.* New York: Oxford University Press, 1959.

Morgan, R. *Sisterhood is powerful: An anthology of writing from women's liberation movement.* New York: Random House, 1970.

Neisser, U. *Cognitive psychology.* New York: Appleton Century Crofts, 1967.

Neisser, U. *Cognition and reality.* San Francisco: Freeman, 1976.

Newell, A., & Simon, H. A. Computer simulation of human thinking. *Science,* 1961, *134,* 2011–2017.

Oakley, A. A case of maternity: Paradigms of women as maternity cases. *Science—Journal of Women in Culture and Society.* 1979, *4,* 607–631.

Osgood, C. E., Suci, G., & Tannenbaum, T. H. *The measurement of meaning.* Urbana, Ill.: University of Illinois Press, 1957.

Peirce, C. S. *Collected papers* (Vol. 5): *Pragmatism and pragmaticism.* Cambridge, Mass.: Belknap, 1934.

Petit, P. *The concept of structuralism: A critical analysis.* Berkeley: University of California Press, 1975.

Piaget, J. *The origins of intelligence in children.* New York: Norton, 1963.

Piaget, J. *Insight and illusions of philosophy.* New York: World Publishing Co., 1971.

Piore, M. P. In R. C. Edwards, M. Riech, & D. M. Gordon (Eds.), *Notes for a theory of labor stratisfactions.* Boston: Health, 1973.

Pirandello, L. Six characters in search of an author. In E. Bentley (Ed.), *Naked Masks.* New York: Dutton, 1952.

Polanyi, M. *Personal knowledge.* Chicago: University of Chicago Press, 1968.

Radnitzky, G. *Contemporary schools of medi-science.* Chicago: Regnery 1973.

Richards, M. *The integration of a child into a social world.* London: Cambridge University Press, 1974.

Rickman, H. P. Vico and Dilthey's methodology of the human studies. In G. Tagliacozzo (Ed.), *Giambattista Vico: An international symposium.* Baltimore: John Hopkins Press, 1969.

Ricoeur, P. *Freud and philosophy: An essay on interpretation.* New Haven, Conn.: Yale University Press, 1970.

Ricoeur, P. The model of a text. *Social Research,* 1971, 38, 529–555.

Ricoeur, P. The hermeneutical function of distanciation. *Philosophy Today,* 1973, 17, 129–143.

Ricoeur, P. Explanation and understanding: On some remarkable connections among the theory of text, theory of action and theory of history. In C. Reagan & D. Stewart (Eds.), *The philsophy of Paul Ricoeur.* Boston: Beacon, 1978.

Riegel, K. *Psychology mon amour: A countertext.* Boston: Houghton, Mifflin, 1978.

Ryan, J. Early language development: Towards a communicational analysis. In M. P. M. Richard (Ed.), *The integration of a child into a social world.* London: Cambridge University Press, 1974.

Ryan, W. *Blaming the victim.* New York: Random House, 1971.

Sahlens, M. D. *The use and abuse of biology.* Ann Arbor: University of Michigan Press, 1976.

Sampson, E. Psychology and the American ideal. *Journal of Personality and Social Psychology,* 1977, 35, 767–782.

Sampson, E. Cognitive psychology as ideology. *American Psychologist,* 1981, 36, 730–743.

Sarason, S. B. *The psychological sense of community.* San Francisco: Jossey-Bass, 1974.

Sarason, S. B. *Psychology misdirected.* New York: Free Press, 1981.

Sartre, J. P. *Search for method.* New York: Random House, 1968.

Schafer, R. *A new language for psychoanalysis.* New Haven, Conn.: Yale University Press, 1976.

Scheman, N. *Individualism and the objects of psychology: Feminist perspectives on epistemology, metaphysics, and the philosophy of Simons.* (S. Hardin & M. Hintikko, Eds.). Netherlands: Reidel, 1981.

Sennett, R., & Cobb, J. *The hidden injuries of class.* Vintage Books, New York, 1973.

Sennett, R. *The fall of public man.* New York: Random House, 1976.

Sève, L. *Man in Marxist theory and the psychology of personality.* Atlantic Highlands, N.J.: Humanities Press, 1978.

Shields, S. Functionalism. Darwinism and the psychology of women. *American Psychologist,* 1975, 30, 739–754.

Shotter, J. *Images of man in psychological research.* London: Methuen, 1975.

Skinner, B. Are theories of learning necessary? *Psychological Review,* 1950, 57, 193–216.

Sullivan, E. V. *Piaget in the school curriculum: a critical appraisal.* OISE Monograph, Toronto, 1966.

Sullivan, E. V. *Kohlberg's structuralism: a critical appraisal.* OISE Monograph, Toronto, 1977. (a)

Sullivan, E. V. A study of Kohlberg's structural theory of moral development. *Human Development,* 1977, 20, 252–276. (b)

Sullivan, E. V. Structuralism in a psychological context. In H. K. Betz (Ed.) *Recent approaches to the social sciences.* Calgary, Alberta, Canada: University of Calgary Press, 1980. (a)

Sullivan, E. V. The scandalized child: Children, media and commodity culture. *New Catholic World,* March–April 1980, pp. 65–71. (b)

Sullivan, E. V. The career and legacy of Jean Piaget. *Queen's Quarterly,* Summer 1981, pp. 341–352.

Takanishi, R. Childhood as a social issue: Historical roots of contemporary child advocacy movement. *Journal of Social Issues,* 1978, 34, 8–28.

Taylor, C. *Explanation of social behaviour.* London: Routledge, 1964.

Taylor, C. Interpretation and the sciences of man. *Reivew of Metaphysics,* 1971, p. 51.

Te Selle, S. *Speaking in parables: a study of metaphors and theology.* Philadelphia: Fortress, 1975.

Touraine, A. Toward a sociology of action. In A. Giddens (Ed.), *Positivism and sociology.* Toronto: Heinmann, 1974.

Touraine, A. Eight ways to eliminate the sociobiology of action. In J. W. Freiberg (Ed.), *Critical sociology.* New York: Irvington, 1979.

Trevarthen, C. The foundations of intersubjectivity: Development of interpersonal and cooperative understanding in infants. In D. Olson (Ed.), *The Social Foundations of Language and Thought* New York: Norton, 1980.

Unger, R. M. *Knowledge and politics.* New York: Free Press, 1975.

Unger, R. M. *Law and modern society: A criticism of social theory.* New York: Free Press, 1976.

Venn, A. & Walkerdine, V. The acquisition and production of knowledge. *Ideology and Consciousness,* Spring 1978, pp. 67–94.

Vico, G. *The new science of Giambattista Vico.* Ithaca, N.Y.: Cornell University Press, 1970. (Originally published 1725.)

Von Wright, G. H. *Explanation and understanding.* Ithaca, N.Y.: Cornell University Press, 1971.

Weil, S. *The need for roots.* New York: Harper & Row, 1971.

Whitteck, C. Theories of sex differences. In C. C. Gould & M. W. Wartofsky (Eds.), *Women and philosophy: Toward a theory of liberation.* New York, Putnam, 1976.

Wilden, A. *Structure and transformation* (Vol. 3): *Developmental and historical aspects* (C. Riegel & G. C. Rosenthal, Eds.). New York: Wiley, 1975.

Wilden, T. *The imaginary Canadian.* Vancouver: Pulp Press, 1980.

Williams, R. Base and superstructure in Marxist cultural theory. *New Left Review,* December 1973, 82, 3–16. (Reprinted in Dale *et al., Schooling and capitalism: A sociological reader.* London: Routledge, 1976.) (a)

Williams, R. *Key words: A vocabulary of culture and society.* London: Fontana Croom Helm, 1976. (b)

Willis, P. *Learning to labour.* Hampshire, England: Teakfield Limited, 1977.

Wilson, E. O. *Sociobiology: The new synthesis.* Cambridge, Mass.: Harvard University Press, 1975.

Wine, J. D., Moses, B., & Smye, M. Female superiority in sex difference competence comparisons: A review of the literature. In C. Stark-Adamac (Ed.), *Sex roles: Origins, influences, and implications.* Montreal Eden Press, 1980.

Winter, G. *Elements for a social ethic.* New York: Macmillan, 1966.

Wishy, B. *The child and the repubic.* Philadelphia: University of Pennsylvania Press, 1968.

Wolff, R. P. There is nobody here but us persons. In C. C. Gould & M. W. Wartofsky (Eds.), *Women in philosophy: Toward a theory of liberation.* New York: Putnam, 1976.

Wright, E. O. Class boundaries in advanced capitalist societies. *New Left Review,* July–August 1976, pp. 3–41.

Zuniga, R. B. *The experimenting society and radical social reform. American Psychologist,* 1975, 30(2), 99–115.

INDEX

Abstract universality, 133–136
Active structures, human action and, 37–38
Adequate account
  as argument, 146, 168
  comparisons of, 166–173
  critical nature of, 149, 171
  as emancipatory praxis, 147–148, 170
  negotiated, 167
Adultocentric interpretation, 121
Age, class dynamics and, 106–108
Agency, structure and, 128–130
Agency sense, in nondominant classes, 81
Agent
  intentional project of, 123
  interventional capacity of, 126
Agent–patient dialectic, 45, 116
Agent–structure dynamics, 141
Age-related institution, legitimacy of, 107
Aggression, vs. violence, 152
Aggro concept, 152–156
Agnosticism, methodological, 20
Ambiguous expressions, 19, 21
American psychology. *See also* Psychology
  as administrative science, 4
  new vocabulary of, 11
  Piaget's influence on, 9
Analysis, instrument rationality and, 4–6
Assimilation, 2
Atomism, of Descartes, 20

Behaviorism, as science, 5–6, 9, 26
Bias, as interpretative horizon, 143
Biological metaphor, in gender dominance, 100
Black nationalism, 90. *See also* Watts youth
Black Underclass, The (Glasgow), 91

Bourgeois hegemony, 90
Brute datum, subjective interpretation at level of, 22

Capitalism
  gender, class, and reproduction of social relations under, 101–105
  habitus of, 70
  nuclear family and, 103–104
  patriarchy under, 99–100
Caretaker–child interaction, 56
Cartesian legacy, 22
Cartesian mind–body dialectic, 21
Cartesian synthesis, Vico's criticism of, 23
Cause, intention and, 141
Child development, environment of, 55
Children
  psychohistory of, 107
  socialization of, 101
Class
  as cultural form, 88
  defined, 78
  as structure of domination, 77–88
Class dynamics
  age and, 106–108
  ethnicity and, 91–99
Classical liberalism, synthesis of, 52
Class society, differential dispositions in, 64
Class structure
  formation and transformative dynamics in, 83–87
  nature of, 83
Class-structured society, personal world in, 65
Cognition, structural conceptions in, 37
Cognitive development, Piaget's study of, 11

189

Communication
  dialogical model of, 58
  gesture in, 57
  individuation and, 56
  personal world as metaphor of, 59–60
Communicative acts, 33
Community economic activity, as enabling
  structure, 140
Computer simulation, psychological theory
  and, 8–9
Concrete universality, 135
Confirmation, in cultural processes, 165
Confrontation, between agents of different
  classes, 64
Consciousness
  human agent and, 39–40
  individuation and, 56
  intentionality and, 40
  in relational totality, 56
Critical emancipating social sciences, 28–29
Critical interpretation, 123–141. See also
  Interpretation
  defined, 123
Critical psychology. See also Psychology
  hermeneutical interpretation in, 123
  resistance in, 139
Critical theory of society, emancipatory psy-
  chology and, 123–130
Critique of Practical Reason (Kant), 27
Critique of Pure Reason (Kant), 27
Cultural forms
  class and, 88, 160–162
  racial characteristics and, 91–92
Cultural science, personal world and, 24
Cultural world, critical interpretation of, 25

Darwin on Man (Gruber), 177
Detached observer, social scientist as, 7
Determinism, defined, 70–71
Determinism–freedom dialectic, 70–71
Developmental psychology, 106–108. See
  also Psychology
Dialectic, of question and answers, 119
Dialectical interpretation, 118–119
Dialectical logic, 20
Dialogue, relationship and, 17. See also
  Communication
Discourse on Method (Descartes), 20
Dislocation, in cultural system, 165–166

Dominant class, 79–81
  capitalism and, 80
  cultural project and, 79
  resistance to, 81
Dominant culture, hegemony and, 89–91
Domination
  class as structure in, 77–88
  cultural forms of, 77
  gender and, 99–106
  hegemony and, 89
  sexism and racism in, 78
  structures in, 76–77
Double hermeneutic, psychological interpretation as, 114–115. See also
  Hermeneutics
Dynamic intentional self, 59
Dynamics–form tensions, 58

Emancipation, cognitive interest in, 30
Emancipatory psychology
  and critical theory of society, 125–130
  freedom and, 139
Empirical–analytic sciences, 28–29
Empiricism, interpretation and, 21–22
Entitlement, 64
Epistemology, shifts in, 24
Ethnicity
  class dynamics and, 91–99
  gender domination and, 105
  in psychology, 137–138
Ethnography
  analysis and, 162–164
  defined, 158
  oppositional culture and, 158–159
Evolutionary theory, male-dominant nature
  of, 134
Exercise of capacity, by agent, 127
Expert, horizon of, 131–132
Explanation
  basic schemes of, 21
  defined, 4–5
Explanation and Understanding (Von
  Wright), 141
Expression, 33–47
  defined, 33
  human. See Human expression

Fact–value dimension, normative element
  and, 6

False Promises (Aronowitz), 178
Family
  under capitalism, 101
  nuclear, 104
  vs. work life, 101
Feminist psychology, 135. *See also* Women
First-person accounts, 151
Football grounds, in *The Rules of Disorder*, 153–156
Form/content distinction, 37
Freedom, 175. *See also* Human freedom
  and emancipatory psychology, 139
  failure of, 175
  vs. order, 174
  structure of, 45
French Revolution, individualism and, 52
Functionalism, 12–17
  intelligence testing and, 12–14
  sociobiology and, 14–17

Gamesman, in dominant class, 80
Gamesman, The (Maccoby), 80
Gender, in psychology, 133–137. *See also* Sexism
Gender domination, 99–106
  ethnicity and, 105
Generalized other, Mead on, 60
Gesture, in communication, 57

Habits
  intentions and, 47–48
  unconscious and, 47–48
Habitus
  defined, 64
  as generative concept, 64
  or order–change relationship, 67
Habitus–project dialectic, 61–73, 77, 169
Hegemonic patriarchy, 102–103
Hegemony
  bourgeois, 90
  domination in, 89
  oppositional forms within, 90–91
  social totality and, 89
Hermeneutical circle, 21
Hermeneutical interpretation, vs. critical, 123
Hermeneutics, science of, 117
Hermeneutic spiral, 116–121
Hidden Injuries of Class, The (Sennett and Cobb), 82, 178

Hierarchy, Descartes on, 20
Historical hermeneutic sciences, 28–29
Historical–hermeneutic shift, 24
Hooligans, in *The Rules of Disorder*, 154
Horizon
  dialectical differences of, 145
  genetic differences in, 144–145
  multiple, 144–150
  perspective in, 144
Horizon problem, in psychological systems of interpretation, 143–144
Human action
  as constitutive act, 63
  critical interpretation of, 124
  dynamic nature of, 120
  intelligibility and, 112
  intentionality and, 40–41, 69
  intention and, 41–44
  personal powers and, 75
  prospective action in, 63
  reasons for, 42
  recursive meaning and, 120
  responsibility and, 44–46
  significance of, 46–47
  social conditions and, 124
  submeanings in, 117
  use of term, 39
Human agency, structure in, 128–130
Human agent, 39–40. *See also* Agent
Human behavior, machines as analogues of, 3
Human expression
  intelligibility of, 34–47
  as relational act, 33–49
Human freedom, self and, 46. *See also* Freedom
Human life, agent and patient concepts in, 126
Human person
  as organizing power, 36
  as "past" or "present," 69
Human projects, limited scope of, 148
Humans, as "complex robots," 3

Identity, individuation and, 64–66
Ideological confirmation, 165
Ideological symbols, utopian symbols and, 69–70
I–it relation, 116

Individual
    in communitarian conception, 53
    defined, 52
    as dominant cultural focus, 131
    as impediment in personal world, 51–52
    liberal ideology and, 52, 136
    male dominance and bias in, 136
    psychology as embodiment of, 54
Individuation
    cultural pattern in, 66
    identity and, 64
Induction–deduction procedures, 20
Instrumental rationality, analysis and, 4–7
Intelligence
    defined, 12
    as dynamic transformational activity, 133
Intelligence testing
    ethnicity and, 137
    functionalism in, 12–14
Intelligibility
    defined, 34
    human action and, 112
    modes of. See Modes of intelligibility
Intensive design, defined, 151
Intention
    cause and, 141
    human action and, 41–44
    teleological, 42–43
    unconscious and, 48
Intentional intervention, 126–127, 149
Intentionality, consciousness and, 40
Intentional self, 68
    social self and, 62–63
Internal interlocutor, 166
Interpretation
    adequacy of, 143–175
    adultocentric, 121
    complementary, 118
    conflicts of, 21
    correct, 118–119
    dialectical horizons of, 118–119
    as dialectic of analysis and synthesis,
        116–117
    genetic, 118
    horizon problem in, 143–144
    multiple horizons in, 144–145
    as normative endeavor, 26
    psychological. See Psychological
        interpretation
    reflexive, 120–121, 138–141

Interpretation (cont.)
    relational quality and, 116
    validation and, 146–147
Interpretative explanation, dialectic of dis-
    tance and relation in, 114
Interpretative inquiry, as method of social
    interpretation, 113
Interpreter, bias of, 143
Interpretive spiral, 116–121
Intervention, intentional, 126–127
IQ testing
    bias in, 13
    social order and, 131
I–thou relation, 55–57, 116

Knowledge, objective, 6
Knowledge and Human Interests (Habermas),
    28
Knowledge-constitutive interests
    concept of, 28
    Habermas on, 125

Language, vs. subjectivity, 19
Language development, Chomsky's concep-
    tions of, 11
Learning theories, assumptions in, 34–35
Learning to Labour (Willis), 108, 156–166,
    172
Let Us Now Praise Famous Men (Agee), 177
Liberal ideology, 136
Liberalism
    classical, 52
    individual and, 52
Life of the Self, The (Lifton), 179
Limitation, in agency–structure interplay,
    162
Linguistic structuralism, 10
Located cultural form, personal world as,
    72–73
Logical analysis, 21
Logical empiricism
    liberalism and, 9
    in societal matrix, 4
Logic of situation, vs. logic of cause, 112

Machines, as analogues for understanding
    human behavior, 3
Mainstream psychology, 1. See also
    Psychology
Males, as dominant class, 79–80

Marxism
    dominance and class analysis in, 77–78
    power relations in, 69–70
Masculinity, labor power and, 164
Master–slave relationship, 76
Maternal deprivation studies, 56
Mechanical metaphors, 3–8
    inadequacies of, 16
    mind and, 11
Median "I" and "me," 62
Mediation, psychological interpretation and,
    113–116
"Messing about," in classroom, 152
Metaphor
    defined, 2
    mechanical, 3–8, 11, 16
    organic, 8–17
    personal, 17–30, 33, 75–76
    praxeology, 3
Metaphors of understanding, 1–31
    mechanical metaphor in, 3–8
    organic metaphor in, 8–17
    personal metaphor in, 17–30
Methodological agnosticism, 20
Milgram experiment, 6
Mind–body dualism, 21
Modes of intelligibility, 34
    mechanical, 34–36
    organic, 36–38
    personal, 38–47
Molecular theoretical system, 6
Moral careers, 153–154, 167
Moral development, in American psychol-
    ogy, 11
Multiple horizons, 144–150. See also
    Horizon

New Science (Vico), 22
Nondominant class, 81–88
    agency sense in, 81
Normative science, 23–24
Nuclear family, capitalist ideology and, 104

Objective knowledge, behavioral psychology
    and, 6
Oppositional culture, 158–160
Oppression, historical projects and, 148
Order–change relationship, 67–68
Organic metaphor, 8–17
    defined, 8

Organic metaphor (cont.)
    in human development, 16
    inadequacies of, 16

Partial penetration, ethnography and, 163
Past–future relationship, 69
Patient, as agent, 116
Patient concepts, in human life description,
    126
Patienthood, 46
Patriarchy
    and domination of women's projects,
        100–101
    as gender domination, 99
    hegemonic, 102–103
    women's movement and, 105–106
Penetration
    in agency–structure interplay, 162
    partial, 163
Performance model, intention and, 150
Person, cultural form and, 73
Personal agency, power and, 75
Personal identity, in class terms, 64–66
Personality
    dialectical conception of, 54
    field theory of, 54
Personal metaphor, 17–30
    human act and, 33, 75
    power and, 75–76
Personal world, 51–74
    boundaries of, 73
    in class-structured society, 65
    conception of, 17–30
    cultural science and, 24–25
    Descartes's doubt and, 20–21
    dominance and, 89
    hegemony in, 89
    human action and, 39–40
    individualism and, 51–54
    integrative components in, 60–73
    integrative polarities of, 61, 128
    interpretation in, 19–20
    as located cultural form, 72–73
    as metaphor of communication, 57–60
    as relational totality, 55–57, 60
    and socialization in tension with social
        transformation, 61–71
    speech in, 55–56
    as text, 111–113
Persons-in-relation concept, 18

Perspective, horizon and, 144
Power
    personal metaphor and, 75–76
    structural relationships in, 75
Powerlessness, depersonalization and, 76
Preconscious–conscious relationship, 68–69
Privileged Ones, The (Coles), 178
Psychohistory, of children, 107
Psycholinguistics, 9
Psychological functionalism, 12–14
Psychological interpretation. *See also*
    Interpretation
    as double hermeneutic, 114–116
    paradoxes of, 113–116
    redundancy and, 118
    task of, 113–116
Psychological sciences, types of, 28
Psychological theory, culture-contextual na-
    ture of, 8
Psychology
    American. *See* American psychology
    as complementary horizon, 132
    critical interpretative, 124
    defined, 1, 67
    developmental, 106–108
    as disciplined inquiry, 132
    dynamic intentional self in, 59
    emancipatory, 125–130, 139
    epistemology and, 24
    ethnicity in, 137–138
    expert and, 131–132
    feminist, 135
    gender and, 133–137
    individualism and, 54
    knowledge-constitutive interests and,
        125–127
    mainstream, 1
    philosophical predispositions in, 27
    psychometric, 26
    reflexive, 7
    social class and, 132–133
    in social order, 130–141
    voluntaristic–intentional theories of, 59
Psychometric psychology, 26

Quasi causal term, in structural determin-
    ism, 141
Question-and-answer dialectic, 119–120

Racism
    antiquity of, 91

Racism (*cont.*)
    cultural forms of, 91
    dominance and, 78
    in working class division, 164
Rationalism, hermeneutical circle and, 21
Rationality
    individualism and, 53
    instrumental, 4–7
Recursive meaning, 120–121
Reflexive interpretation, 120–121
    need for, 133–141
Reflexive psychology, 7
Relational act, expression and, 33
Relational totality
    consciousness in, 56
    personal world as, 55–57
Reproduction, class structure and, 108
*Reproduction in Education, Society and Cul-
    ture* (Passeron), 169
Researcher, credibility of, 168
Resistance
    as agent process, 139
    community consensus and, 140
    dominant class and, 81
    of English working-class youths, 82
    of lower-class adolescent, 82
Responsible action, 44–56
Responsible agent, 45–46
Resymbolization
    example of, 115–116
    and problems of horizons, 118–119
    social scientific accounts as, 149
Resymbolizer, Freud as, 115
Rowdies, in English school system, 154
Rules of Disorder, The (Marsh *et al.*),
    150–157, 159, 167–168, 170–171, 174

Sane Asylum (Hampden-Turner), 179
Schoolboys, as forms of life, 160
Schooling in Capitalist America (Bowles and
    Gintis), 169
Schoolroom setting, 152
Self. *See also* Social self
    dynamic intentional, 59
    intentional aspect of, 68
Self as agent, 39
    human freedom and, 46
Sex differentiation, role and status related to, 99
Sexism
    cultural penetration in, 163
    dominance and, 78

Situation, logic of, 112
Six Characters in Search of an Author (Pirandello), 7
Sociability, cognitive and moral aspects of, 56
Social class, psychology of, 132–133
Social contract, individual and, 52
Social investigation, interpretative inquiry as method of, 113
Socialization
 determinism and, 71
 social transformation and, 61–71
Social order
 freedom and, 174
 maintenance of, 71
 psychology in, 130–141
Social sciences
 as branch of evolutionary biology, 14
 critical emancipatory, 28–29
Social scientist, detached observer role of, 7
Social self. *See also* Self
 intentional self and, 62–63
 Mead on, 60
 social environment and, 67
Social totality, hegemony in, 89
Social transformation, socialization and, 61–71
Society
 class-structured, 65
 critical theory of, 125–130
Sociobiology
 criticism of, 15
 functionalism in, 14–17
Sociobiology (Wilson), 14
Solidarity, totalitarianism and, 175
Sons of Watts Improvement Association, 93, 98
Speech, in personal world, 55. *See also* Language
Speech codes, in class-structured society, 65
State of nature, Hobbes on, 52
Structuralism
 beginnings of, 9
 defined, 10
 linguistic, 10
 organism and, 12
 power of, 11
 social science theories of, 25–26
Structure of domination, 130
Structure of freedom, 130
Structure of Scientific Revolutions (Kuhn), 2

Subjectivity, vs. language, 19
Success, fear of, 13
Symbolic interaction theory, Mead on, 59–60

Tabula rasa concept, 22
Text, personal world as, 111–113
Theory language, ideological, 2
Totalitarianism, solidarity and, 175
Totality
 relational, 55–57
 in structuralism, 10
Town boys, aggression among in English schools, 154
Transaction, reflexive interpretation and, 121
Truth, absolute vs. relative, 145
Truth and Method (Gadamer), 23

Unconscious
 as generative structure, 68
 habits and, 47–49
 intentions in, 48
 metaphors of, 1–31
Universality, abstract vs. concrete, 133–136
Utopian symbols, ideological symbols and, 69

Validation, interpretation and, 146–147
Value neutrality, as ideology, 25
Viconian legacy, 23
Violence, aggression and, 151–152

Watts riots, Los Angeles, 92–98
Watts youth, schematic interpretation of before and after riots, 94–97
Women
 concrete universality and, 136
 as "incomplete men," 99
 in occupational ghettos, 104
 oppression of under hegemonic patriarchy, 102–103
 as private sphere, 101
Women's movement, patriarch and, 105–106
Working-class youth, conformist, 161
Work life, vs. family, 101
World of Pain: Life in the Working-Class Family (Rubin), 178

Young Radicals (Kenniston), 67
Youth culture, 90